T0136820

Toby Carlson is a professor of meteorology, Emeritus, at Penn State, where he has taught a variety of courses for over 30 years. He has also published a large number of research papers and a book titled *Mid-Latitude Weather Systems*.

Paul Knight is an instructor in the Department of Meteorology at Penn State. He teaches courses on weather forecasting and weather applications, is an executive producer of the meteorology department's daily TV weather show (*Weather World*), and also serves as state climatologist for Pennsylvania.

Celia Millington Wyckoff was an editor for the *World Campus* at Penn State. She is also a well-known musician in the State College area.

AN OBSERVER'S GUIDE
to
CLOUDS AND WEATHER

A NORTHEASTERN PRIMER ON PREDICTION

TOBY CARLSON, PAUL KNIGHT, AND CELIA WYCKOFF

AMERICAN METEOROLOGICAL SOCIETY

An Observer's Guide to Clouds and Weather: A Northeastern Primer on Prediction © 2014 by Toby Carlson, Paul Knight, and Celia Wyckoff. All rights reserved. Permission to use figures, tables, and brief excerpts from this book in scientific and educational works is hereby granted provided the source is acknowledged.

Front cover photograph by Jerry Wyckoff.

Published by the American Meteorological Society
45 Beacon Street, Boston, Massachusetts 02108

The mission of the American Meteorological Society is to advance the atmospheric and related sciences, technologies, applications, and services for the benefit of society. Founded in 1919, the AMS has a membership of more than 13,000 and represents the premier scientific and professional society serving the atmospheric and related sciences. Additional information regarding society activities and membership can be found at www.ametsoc.org.

Library of Congress Cataloging-in-Publication Data

Carlson, Toby N.
  An observer's guide to clouds and weather : a northeastern primer on prediction / Toby Carlson, Paul Knight, and Celia Wyckoff. — First edition.
      pages cm
  Summary: "A basic introduction to making weather predictions through understanding cloud types and sky formations"—Provided by publisher.
  ISBN 978-1-935704-58-4 (pbk.)
  1. Cloud forecasting. 2. Weather forecasting.
I. Knight, Paul. II. Wyckoff, Celia. III. Title.
  QC921.C34 2014
  551.63—dc23
                    2014033512

CONTENTS

PREFACE
Including some information about the authors' backgrounds
Toby N. Carlson, Paul Knight, and Celia Millington Wyckoff

As a young boy, I was an enthusiastic sky watcher and weather buff, despite my nearly total lack of knowledge (or interest) in the physics and mathematics governing the atmosphere. When asked by Professor Stapleton, a meteorologist on the faculty of the University of Massachusetts whom I visited at the age of ten with my Mother and who was helping me design a rain gage from a coffee tin, whether I knew what *circumference* was, I blandly admitted that I did not, though I thought that the subject would be taken up later in the fourth grade. Knowing nothing about science or mathematics, I was nevertheless an enthusiastic sky watcher and amateur meteorologist from an early age. Whether it was looking at fair-weather cumulus clouds for the shapes of animals or faces, or searching the horizon for the telltale harbingers of a snowstorm, I was excited by whatever I could see.

At about that time, my parents gave me a thin, blue hardbound book titled *Weather*, by Gayle Pickwell. Inside were numerous photographs of clouds with captions that included mention of their significance and what they foretold for changing weather. The clouds all looked familiar to me, but I was thrilled to finally learn what they meant. I became a sort of family guru of the weather. I kept a weather diary in which I dutifully wrote my forecasts. Members of the family frequently asked me for a forecast, and I delighted in providing such a service, claiming that I was 80% accurate—though I did not know what that meant, as I had only the vaguest idea of what constitutes a percentage. My method of scoring would have baffled a statistician. Yet, it was fun to inform and even amaze my relatives that a seemingly violent snow squall was not a real snowstorm but a passing snow shower that would end in a few minutes.

My technical knowledge was limited and my available hardware primitive. I bought a barometer from a local jeweler, a young fellow just back from WWII, who must have regretted the sale, as I continued to pester him almost on a daily basis and for months afterward in order to examine the weather map published in his daily copy of the *New York Herald Tribune*; it was the first of its type I'd ever seen. Today, that same barometer hangs in a room of my house. Later, I would query the U.S. Government Printing Office for a subscription of their weather maps, which would arrive by mail approximately 36 or 48 hours after map time. How exciting it was to see the symbols depicting weather, clouds, and winds in exotic places such as Helena, Montana, or Edmonton, Canada, albeit two days old. A cousin gave me a recording wind anemometer for my birthday. I don't recall who had taken the risk of mounting the device on our roof.

For a wind vane I had mounted a crude wooden arrow impaled on a nail and stuck in a tree barely above eye level between our house and garage. Miraculously, the device seemed to work well. I also had a rain gage sitting on the railing of my back porch (a coffee can) and a thermometer nailed to the house aside my back door. I kept a faithful weather log and eventually acquired a meager collection of weather books.

When I thought to write this book, I approached my colleague, Paul Knight, with the idea of collaborating in the writing. Paul is a staff member and instructor in the Department of Meteorology at Penn State. Once a student in one of my classes, Paul had evolved into one of the most knowledgeable meteorologists I know, a weather enthusiast like myself, and an expert on all sorts of weather phenomena, especially those specific to this region. Paul is best at telling his story in his own words:

> Like my colleague, Toby Carlson, I also have a long-standing interest in observing and predicting the weather that predates my knowledge of any atmospheric science. From walking home from grade school during a hurricane to volunteering to shovel the sidewalks so I could be outside during a winter storm, my recollections of growing up are peppered with accounts of all sorts of storms. To the amazement of my older brothers, I asked for and received an aneroid barometer for my 10th birthday. My father risked his neck to mount

an anemometer on our roof so I could measure the wind direction and speed when I turned 15. I had scanned or read most of the books about weather in the local library and was called on in various grades in primary and secondary school to give the weather forecast. Of course, all those predictions were just parroting what I had heard or read earlier that day—with a little bit of personal forecast experience and some wishful thinking.

Celia Millington Wyckoff has been a friend of mine for many years. Her skills as an editor are essential to combat my occasional sloppiness, and her lack of knowledge of meteorology made her a perfect sounding board for improving the clarity of my explanations. I began my correspondence with her remarkable father, a published author of a geological textbook and an avid cloud photographer, on the subject of the names and the significance of the clouds in his photographs. I regret never having met Gerry Wyckoff in person, but when he died I suggested to Celia that we publish a book of his photographs and the interpretation of the cloud forms depicted in them. The idea languished for several years until a series of unrelated events caused us to revive the idea in a greatly expanded form, which became this book. Gerry Wyckoff's namesake appears as a cartoon character in the first chapter of this book. Celia describes our collaboration:

> Sometime in or around 2003, my dad called to ask me if I knew any meteorologists at Penn State who might be interested in owning his numerous slides of clouds. I recognized that my dad, who had been a nature photographer in retirement, was beginning to divest himself of his possessions at the age of 92.
>
> Now, Penn State is a big place, and I—a writer/editor in another part of the university—had never worked with the College of Earth and Mineral Sciences. But it just so happened that I did know a meteorologist, through a music connection. That meteorologist was Toby Carlson, a faculty member in meteorology at Penn State.
>
> I wasn't optimistic that he'd be interested in the slides—after all, I was aware that these days weather forecasting is accomplished through computer modeling. To my surprise, Toby said he would like to see my father's slides. At that point, I gave Dad and Toby each other's contact information, and stepped out of the picture. What ensued was an e-mail friendship that lasted for several years, until my dad became too feeble to maintain communication. Within that period, Toby was able to use some of the slides in a class he was teaching. My father's mission had been accomplished—his cloud slides had found a caring home. My only regret was that Dad and Toby never got

to meet face to face. Yet this contact served as an inspiration and the germ for this book. Many of the cloud photographs in this book were taken by my father. After Dad and Toby had struck up a friendship, my father told me about a remark Toby had made that really stuck with him: "Nobody forecasts the weather by looking at the sky anymore." This book is our opportunity to change that, and to study the clouds that will enable one to read the sky.

The first chapter presents highly simplified explanations of the physical processes that govern the behavior of the weather without recourse to complex physics. The second chapter relates these processes to classic models of weather patterns, which allow one to integrate observations into a unified picture in the form of a weather map. The third chapter presents a compendium of cloud pictures discussed in terms of their significance, placing them within the schematic weather pattern treated in the first two chapters. Each of the main cloud types is presented within the context of the classic weather pattern, revealing a clear picture of where the observations fit within the weather map, and enabling the reader to understand the weather maps one might find on websites or in newspapers. Chapter 4 discusses further aspects of cyclogenesis as well as some small-scale weather systems that do not fit the classical mold. Chapter 5 deals with some aspects of modern weather forecasting, and Chapter 6 addresses the needs of the modern amateur weather forecaster and observer, stressing simple observations that can be made from one's home.

I will acknowledge up front the geographically parochial nature of the examples presented in this book. These are heavily weighted toward weather patterns that one normally experiences in the northeastern United States, for which Pennsylvania might serve as an iconic representation of that part of the country. Each region has its own peculiar weather and no book or set of principles could possibly present a completely consistent paradigm governing the weather for every area. Indeed, even within a specified region, smaller-scale climate regimes exist and within these regimes yet smaller-scale regimes can be found, and so on. Were we to address the weather patterns and weather vicissitudes characteristic of all parts of the country, this book would be many times its present size and would require an intimate familiarity with these other climate regimes that none of us now possesses. Of course, the principles discussed in the first three chapters are applicable anywhere in the world, as are the significance of the various cloud forms discussed in Chapter 3.

Yet, given the infinite variations of weather on smaller and smaller scales, the northeastern part of the United States does have its signatory weather patterns that have predictive power. Clouds and weather patterns discussed in the book are so frequently observed in relation to one another that once the patterns are recognized the observer can easily interpret their meaning and make predictions that will verify most of the time correctly, at least for the subsequent 24–48 hours.

This book is not intended to be a formal text book in which a principle or idea is introduced and then the discourse moves on to the next item. Rather, it is written in an informal style, intended to draw the reader in. While individual chapters can, to some extent, stand alone, all chapters relate to each other as an integrated whole. For that reason, ideas and principles are often repeated in different ways in different chapters and even in the same chapter. I believe that such redundancy is important for learning. I can only judge this by reflecting on my younger self and what I would have wanted to read about the weather before I knew anything about the physics and mathematics behind the phenomena that so intrigued me.

We recognize that, while some meteorological students and professionals might find aspects of this book a bit elementary or oversimplified, the book is intended to clarify and to make very simple some very difficult and complex science for the layman, or even for the beginning meteorology student. Still, we believe that even beginning meteorology students will find much of the insights and information in the book to be very useful in their career.

Most of contemporary weather forecasting has been replaced by high technology, much of which is available to the public. It is currently possible with the aid of a computer to access the latest weather map, satellite photograph, and radar display. It is also possible to access via the Internet forecasts up to 10 days or more in advance and to view the corresponding weather maps generated by powerful computers, which use banks of differential equations and mounds of data to make their predictions. Yet, we, the authors of this book, wish to make the reader's experience much more intimate with the atmosphere than simply viewing computer-generated output.

Despite the reliance on products visible on the computer screen, it is our hope that this book will help refine one's weather observing skills to the point that the reader can make reliable estimates of what today's and tomorrow's weather will be like! To do so, for starters, one need only look out the window. In so doing, we want to encourage the reader of whatever age to view the sky with understanding and delight while still conscious of the basic principles and processes that enhance one's powers of observation. Nothing would delight us more than if the reader were to go beyond our explanations and to discover for him- or herself an expanded understanding beyond that derived from reading this book.

Toby Carlson
January 2013

ACKNOWLEDGMENTS

This book is dedicated to Jerry Wyckoff, whom I never met in person but whose friendship, enthusiasm for the weather, and his many cloud photographs provided the inspiration for its creation. We would also like to express our deep appreciation to the American Meteorological Society and specifically to its director, Keith Seitter, and hard-working editors Ken Heideman, Beth Dayton, and Sarah Jane Shangraw for their technical support and good-natured assistance during and after the book was written. And to Erin Greb, who created many of the drawings.

"A meteorologist is illiterate who can not read the sky."—*Alistair Fraser, Professor of Meteorology, Emeritus, Penn State University*

CHAPTER 1
THE BASIC PROCESSES THAT
CREATE THE WEATHER

Why read the sky?

Humans have always been fascinated by the prospect of knowing the future, and knowing tomorrow's weather is no exception. However, most people today don't pay close attention to weather forecasts unless the predictions are likely to impact or inconvenience their daily lives. Forecasts of canceled school days, rained-out golf games, slippery highways, flooding, tornado or hurricane warnings, and power outages are followed by the public with the anticipation and awe once reserved for medieval soothsayers or Roman auguries, who regarded the innards of chickens in order to foretell the fortunes of individual or military adventures. Today, tranquil weather does not seem to generate such excitement about tomorrow's weather.

Some parts of the United States have placid weather conditions, and it seems no one much cares about weather forecasts. In other parts, such as the northeastern United States, residents tend to take weather rather seriously. Mark Twain joked about how fast New England weather changes. Weather over the northeastern part of the country is situated smack in the middle of the belt of storms, straddled on the east side by the ocean, on the west side by a large land mass, on the north by a vast land mass that becomes exceedingly cold in winter, and on the south by a source of warm, moist air from the Gulf of Mexico. Not surprisingly, the weather in that part of North America is highly changeable.

Yet, despite fascination with the weather, people may not know why the sky and clouds appear as they do, what they portend, or how to simply appreciate what they do. If we look around us, we must realize that half of what we see lies above us in the sky—especially its patterns of clouds. Moreover, as we walk along the street we are unable to see much of the lower half dome, which is the land surface, so obscured by houses, trees, and other obstacles. To ignore more than half of what we see around us limits our understanding of our physical world. Artists, composers, photographers, and poets have always been acutely aware of the beauty of the sky and its panorama of clouds. The great early 19th-century English painter Joseph Turner portrayed clouds in exquisite detail. His German contemporary, Ludwig von Beethoven and 17th-century Englishman Mathew Locke composed music evoking the sound of storms. John Muir, 19th-century explorer and conservationist, photographed the western sky in all its glory. Visual artists, musicians, and writers have made the sky and the weather a subject of their craft.

Aesthetics aside, we will need to set the stage for an understanding of the weather and how one can become one's own weather forecaster. We will first introduce the physical principles that govern the movement, development, and decay of weather systems (highs, lows, and fronts), expressed not in technical terms but in the language of everyday imagery. Once this material is presented we can proceed to Chapter 2 in which the basic features of the weather map and satellite imagery are presented and then related to the principles introduced in the first chapter from which these map features, including the explosive growth of cyclones, can be understood. Chapter 3 presents the principle characters in the sky drama, the clouds. The various types and patterns of clouds that one can observe are shown in numerous photographs and on satellite and radar imagery, which are related to the weather map features introduced in the previous chapter. Chapter 4 introduces some smaller-scale details commonly observed on the weather map and satellite imagery. The last two chapters relate the previous material to weather forecasting. Given this knowledge it should be possible for the keen observer to outperform computer model forecasts for short periods, possibly up to 36 hours or more. The latter is one ultimate goal of this book—not to make professional weather forecasters but to allow the individuals to appreciate, understand, and even anticipate future weather using only their observational skills.

We therefore need to set the stage for the sky drama and introduce the characters so that our readers may know their traits and what they portend.

This will be the subject of Chapter 3. First, however, we must understand why the weather acts as it does by explaining what forces and processes are responsible for the sky drama. What makes things occur as they do? For this we need to look at some basics here and in the next couple of chapters.

Why the winds blow as they do

The atmosphere is a restless heat engine. It is always in motion; it is always a little bit out of balance, as we will describe. Its internal currents of air are continually going somewhere, purposefully and in all directions like a harried behemoth. Its continual lack of balance is due to internal forces that are never quite in equilibrium with one another. While its objective is always to redress this imbalance, the atmosphere never quite achieves a perfect balance. Imagine a tightrope walker as he moves along the rope (Figure 1.1a). We marvel at his sense of balance, but if we look closely we see that the person is constantly shifting weight from one side to another, tipping the horizontal pole up on one side, then down, trying to right a sense of imbalance. So also does the atmosphere strive to correct for its imbalances with its subtle internal forces and the resulting motions.

What do we mean by a force? A force is simply a push or a pull, without which nothing either moves or changes its speed or direction. Imbalance of forces always leads to movement. According to Isaac Newton, a force exerted on a body causes that body not just to move but also to accelerate or decelerate. As an aid in visualizing an imbalance of forces, consider a tug of war between two groups of people. Each group pulls toward its side with great force on a rope in an attempt to move its adversary toward its own side. Each

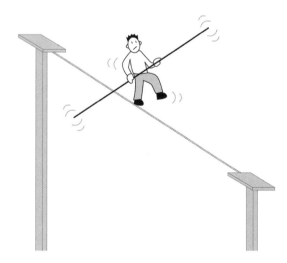

FIGURE 1.1A. A tightrope walker is constantly attempting to maintain balance.

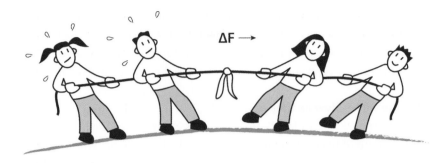

FIGURE 1.1B. Tug-of-war schematic. The pair of individuals tugging on the right is exerting a slightly greater force in its direction than the pair on the left, resulting in a movement toward the right, represented by the symbol ΔF.

side may be exerting a large force on the rope, but if the other side exerts the same force in the opposite direction, no motion is possible. Suppose, however, that one side is just slightly stronger than the other. In that case, a force difference, or imbalance, occurs and the weaker side begins to move forward toward the stronger side.

We illustrate this balance, or lack of balance, in Figure 1.1b. The slightly superior force exerted by the team on the right-hand side over that of the left causes the rope and all those pulling on it to accelerate from zero speed to some motion toward the right side. In fact, the force imbalance is liable to come quickly into equilibrium due to the addition of an opposing force—the drag of the feet of the left-hand team against the ground, not to mention increased resolve by the team on the left.

In the atmosphere, the forces are enormous, but the imbalances, the difference between the big forces, are relatively small compared to the forces themselves (as in the tug-of-war idea), although the imbalances are still large enough to cause storms to intensify, thunderstorms to arise, and the winds to blow, even with great speed. As we will see in later chapters, and as our previous discussion of the tug of war demonstrates, the imbalance of forces in the atmosphere is often quickly restored because of the establishment of processes that counteract the initial imbalance. For example, storms may suddenly intensify, but the process of intensification usually does not proceed unchecked, and eventually—often quickly—the storm reaches some maximum intensity and then begins to weaken. The atmosphere pushes back, so to speak. Were the character shown in Figure 1.1a to lose his balance, he might reach the ground falling at a fatally rapid speed. If a net were strung below the rope, it would keep him from attaining bone-breaking speeds in the event of a fall, allowing him a soft landing. Like the tightrope walker, who has a safety net to keep him from falling all the way to the ground, the atmosphere also does not permit all of its enormous energy to be released during an instability but only a small amount of this pent-up energy is realized before the atmosphere is able to restore balance. The atmosphere pushes back, so to speak, in an attempt to restore balance. Such episodes of dramatic imbalance or instability in atmospheric processes are therefore inherently self-limiting, just as the drag on the ground and the reaction of the weaker side in the tug-of-war example will lead to increased resistance to the team with initially superior strength, and perhaps to a reversal of the force imbalance. We will shortly return to the tug-of-war imagery to illustrate its application to the atmosphere.

The pressure force pushes the air around

Pressure, or atmospheric pressure, is just such an atmospheric force, although as a pressure it must be divided by a unit area (typically the surface area: e.g., square inches) to give it the units of pressure (e.g., in pounds per square inch [psi]). Other, and more customary, units of atmospheric pressure are millibars (mb), or millimeters of mercury. More precisely, atmospheric pressure is the weight of the entire air column on a unit surface; a typical sea level pressure is about 15 psi or about 1 kilogram per square centimeter (kg/cm^2), which is equivalent to the unit of 1 bar. Note that 10 meters (m), about 30 feet, of a water column weighs approximately the same as about 30 in. of mercury, both being a measure of the weight of an entire atmospheric column of air on that same area.

The word *bar* comes from the Greek word *baros*, which means weight—in this case, the weight of a column of air. From this word we also derive the word *barometer*, meaning an instrument to measure the weight of the air; that is, the pressure. Modern weather maps customarily are labeled in units of millibars, where 1000 mb is equal to 1 bar. Inches of mercury is also a popular unit of pressure; that is, 30.15 inches. Thus, one may find lines of constant pressure, called isobars, drawn on a weather map. One finds isobars labeled successively at regular intervals, typically 4 mb apart—for example, 1020, 1024, 1028, the unit of millibars being implicitly understood.

Since air, like water, is a fluid, it exerts its force to the sides of a column of air or water as well as toward the ground. So, if two points on the weather map—say, 500 km apart—have the same atmospheric pressure, no net horizontal force is exerted on the air between the two points. Now let the pressure at one point be somewhat lower than at the other point. A force, called the *pressure gradient force*, will be exerted on the air from the higher pressure toward the lower pressure, just as in the tug-of-war example.

Before we go on, let us define what we mean by a gradient. Anyone who has bicycled up a mountain knows that a gradient of 100 m climb over a horizontal distance of 1,000 m (a 10% *grade*) is a very steep incline. A climb of 10 m over that same distance of 1,000 m (a 1% grade) would be easily mounted. We can speak of a pressure gradient, as illustrated in Figure 1.2, showing a 5-mb difference in pressure between Pittsburgh and Philadelphia, a distance of about 500 km as the crow flies. That pressure difference of 5 mb divided by 500 km, constitutes a pressure gradient force, a push by the air from the higher toward the lower pressure, which is to say that the pressure force would be directed from Pittsburgh toward Philadelphia. That gradient would certainly be stronger than one in which the difference in pressure between the two cities might be only 1 mb; in that case, the pressure gradient would be 5 times weaker. Note, however, that the pressure on weather maps has to be corrected to the same reference height for the pressure gradient to be a true horizontal pressure gradient. This reference level is sea level. However, we will not digress into the subject of how this correction to sea level is done.

All sorts of gradients exist, not just height gradients on an incline or pressure gradients. A difference in temperature between these two locations of 10 degrees Centigrade (or Celsius; abbreviated as °C) would constitute a temperature gradient, although this could not correspond to any force. However, temperature gradients and pressure gradients are intimately associated in the atmosphere and, as such, both figure closely in the movement of air.

Why do pressure, air density, and temperature decrease with increasing altitude?

Returning to pressure gradients, we will digress for a moment to wonder why the pressure at the surface of the atmosphere is so much larger than at, say, an elevation of 10,000 m. Because pressure decreases with height, the pressure gradient between the surface and 10,000 m, in principle, exerts a net force upward—higher toward lower pressure. Yet, the pressure gradient force acting to push upward is almost exactly balanced by the gravitational force, which acts to push the air downward. In fact, the weight of the air is the very reason why the pressure decreases with height; it is due to gravity.

Imagine that 25 pillows stuffed with some compressible material such as foam are piled one on top of the other. The force of gravity pressing down on the lowest pillow would be that of the weight of all the other pillows above it. Similarly the weight on, say, the fifth pillow from the ground would be

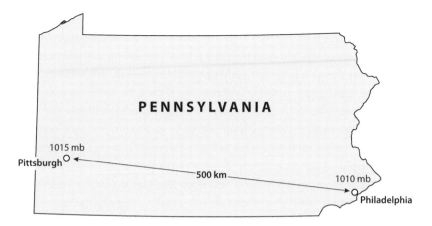

FIGURE 1.2. Illustration of a pressure gradient between Pittsburgh, where the sea level pressure is 1015 mb, and Philadelphia, 500 km from Pittsburgh, where the sea level pressure is 1010 mb.

only 20 pillows. The lowest pillow would be squashed by the weight of the column and would necessarily be compressed into a denser mass than the fifth pillow from the ground. So the lowest pillow would be squeezed tighter by the weight of all those pillows, unlike the ones above, such as those near the top, which would have less weight above them. The pressure (the weight of the pillows) and the density of the pillows would be larger at the surface of this great mound of pillows than at the top. Lower pillows would be compressed more tightly than the upper ones, so that the lower pillows would have a higher density (mass of material per volume of pillow), having the same amount of material packed into a smaller space.

Pressure decreases with height as does the density of the air, whereby the air is compressed more at the surface by the weight of all the air above it than at, say, 10,000 m above the ground where the weight of the air (and its density) would be much less than at the surface. Similarly, the air above 10,000 m is compressed even less, all the way up as far as one can imagine. Even at 100,000 m elevation, the column of air above that level still has some weight, which presses down on the air below.

To give the reader a sense of how pressure changes with altitude, the atmospheric pressure is approximately equal to half that of the surface pressure at approximately 5,500 m above the ground. If you rise another 5,500–11,000 m, the atmospheric pressure is about half that at 5,500 m and about one-quarter that at the ground. If you rise another 5,500 m, the atmospheric pressure decreases to about one-eighth that at the ground. And so forth up to heights so great that atmospheric pressure no longer has any significance as the atmosphere then consists of just individual molecules of oxygen, nitrogen, and other constituents just bopping around, including charged particles.

So the air exerts a force not only upward, because of the pressure force decreasing with height, but also horizontally. If pressure were the only force to consider, air would always flow horizontally from high to low pressure, which is to say from regions of high pressure on a weather map toward regions of low pressure. If this were actually the case, however, low pressure systems would fill up in no time and the pressure gradients across a weather map would be eradicated; clearly this does not happen in the real world. Since low or high pressure systems do not readily disappear, we must look for another force to counter the pressure gradient force, just as gravity must balance the upward pressure gradient force in the vertical.

Why temperature decreases with height is not obvious. The ancients thought that temperature increased with height. In Greek mythology, Icarus fashioned a pair of wings consisting of feathers stuck together with wax, but he attempted to fly so high that he caused the sun to melt the wax from his wings. Icarus then fell to his death in the sea, which was thus named the Icarian Sea, the latter situated near the Greek island of Samos. In fact, temperature decreases with height over the lowest 10–15 km, approximately at a rate of about 5°–6°C per kilometer. Like virtually all gasses, air cools as it expands and warms as it compresses, a principle on which rests the workings of our familiar and much employed air conditioner. Expansion or contraction of the air as it rises or descends is caused by the decrease in density with height, thereby requiring air to cool as it ascends and warm as it descends.

Solar energy is mostly absorbed at the ground, with very little of it being absorbed by the air during the passage of the sun's rays through the atmosphere. As a result, the warmest air is typically found near the surface. Heating of the atmosphere takes place primarily through the transfer of heat energy from the surface (sea or land) to the air near the surface and thereupon, by mixing, to the air immediately above the surface, and ultimately to levels much higher in the atmosphere.

Because air is circulating constantly both horizontally and vertically, the result of weather processes soon to be discussed, every bit of air near the surface must have been in the upper atmosphere and every bit of air in the upper atmosphere must have been near the surface at some point during its wanderings. Air reaching the upper levels in the atmosphere (say tens of kilometers) from near the ground will have therefore cooled by some tens of degrees Celsius, thus requiring it to possess a very low temperature at high altitudes. Similarly, air reaching the surface from some high altitude must warm considerably. Density also undergoes the same changes with changing altitude. Heating near the surface and the mixing of air between high and low levels maintain the decrease in temperature with height.

Continuous heating of the surface does not result in a runaway heating of the atmosphere because the earth (and ultimately the atmosphere) is constantly losing heat to space through losses of long wave (thermal) radiation—the kind of radiation one senses across the room from a well-lit fireplace. (Thermal radiation is much in the news because of its sensitivity to absorption by greenhouse gasses—carbon dioxide, water vapor, and methane.) These long wave radiation losses are greatest at the ground and decrease with height. Radiation leaving the earth's atmosphere therefore consists of contributions from all levels in the atmosphere, the largest contribution being from the earth's surface. Averaged over the entire earth's surface, the incoming solar radiation and the outgoing thermal radiation are in close balance, although the distribution of heat with latitude is far from uniform, with the input of solar radiation being much greater than the outgoing thermal radiation at low latitudes than at high latitude, and the output of thermal radiation to space being much greater than the input of solar radiation at high latitudes than at low latitudes. The result is a heat deficit at high latitudes and a temperature gradient between the low and high latitudes. This heat imbalance and its attendant temperature gradient fuel the motion of the airstreams, a process that will be discussed in great detail in the next chapter.

Coriolis force

Now, let's consider a boy named Jerry, who is standing on a merry-go-round rotating counterclockwise (Figure 1.3). He sees his mother on the right of the figure beckoning to him to come to her and he obeys her command. Jerry attempts to walk a straight line in the direction of his mother, but after several steps he finds himself to the right of his initial point on the floor of the merry-go-round. In so doing, Jerry would be aware of a seeming force pulling him to the right, so that he has no choice but to move along the trajectory J_1 to J_2, instead of along the trajectory followed by the seat that Jerry had been sitting in, which moves from points a (also the location J_1) to b. From his standpoint, however, Jerry perceives his motion toward his mother to be along the somewhat *clockwise* curved path from J_1 to point J_2 (the dashed line); the little arrows along the path J_1 to J_2 denote the direction Jerry was moving toward his mother at various instants along his path.

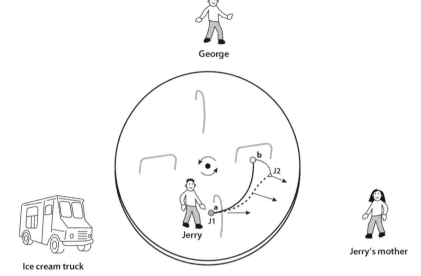

FIGURE 1.3. Illustration of Coriolis force on a merry-go-round that is rotating horizontally counterclockwise. Jerry starts out at point a, also labeled J1, while attempting to walk a straight line toward his mother on the far right. Were he to have remained at his seat at J1, Jerry would have rotated with the merry-go-round to point b. Instead, in trying to walk a straight line, he moves to point J2, to the right of point b. The little arrows denote the directions that Jerry was facing toward his mother at all points along his trajectory as he attempted to walk in her direction. To someone sitting on Jerry's original seat at J1, Jerry would appear as though he were walking a curve from point b to J2 (thin dotted line). Other points referred to in the discussion are the ice cream truck on the left and Jerry's friend George above.

The rightward displacement is caused by the merry-go-round turning counterclockwise (as seen from above), pulling Jerry on a clockwise curve to the right while he is trying to walk a straight line toward his mother. Even if Jerry had decided to annoy his mother and walk in the opposite direction from a direct line toward her (toward the ice cream truck on the left in the figure), he would still feel the same tug to his right. Because the perceived force is the result of the object (Jerry) trying to walk a straight line, this would be true if Jerry's destination were any other location; for example, if he wanted to walk toward his friend George in the figure. Interestingly, if the merry-go-round had been rotating in the opposite sense (clockwise), the tug that Jerry feels would be toward the left in a counterclockwise sense. This apparent force pushing Jerry to the right (or to the left) is called the Coriolis force (named after the man who discovered it in 1835, Gaspard Coriolis). The

Coriolis force therefore pulls air (and water) and objects in flight to the right in the Northern Hemisphere, but toward the left in the Southern Hemisphere because the sense of Earth's rotation, as viewed from above the South Pole, is opposite to that viewed from above the North Pole. Mathematically, the Coriolis force is proportional to Earth's rotation rate and proportional to the speed of the air relative to Earth; an air parcel at rest would therefore experience no Coriolis force.

The Coriolis force is not really an independent force such as gravity, but an artificial force resulting from inertia: Bodies in motion tend to keep a straight line; inertia is what keeps a bicycle upright, as long as it is in motion. Yet in the rotating Earth system we can consider the Coriolis force to be real, as would Jerry in trying to walk a straight line, but instead he follows a curved path to the right of his motion, point a (or J1) to J2.

Imagine further that just after Jerry gets the call from his mother, the merry-go-round operator accidentally leans on the lever controlling the rotation rate and the merry-go-round begins to turn very rapidly. Poor Jerry! By the time he has gone a couple of steps toward his mother, the rotation has moved Jerry around to the opposite side, requiring him to turn back to keep a straight line toward his mother. Jerry experiences an apparent force pushing him to the right, but it is only a virtual force resulting from his attempt to move in a straight line. Likewise, air (and also water) experiences this same apparent force resulting from its attempt to move in a straight line on a rotating Earth, a consequence of *inertia*—things tend to remain in motion (or at rest) until some external force is applied. In this case, no external force is being applied, although it seems to Jerry that he is being pushed to the right. This could cause Jerry to return to the point where he started after one rotation of the merry-go-round, while traversing a path describing a clockwise loop rather than just a curve to the right, as shown in Figure 1.3. In this case, Jerry would never reach his mother, forever walking in a clockwise loop. The same effect would be noticed if Jerry had walked toward his mother at a very slow rate compared to the speed of a point on the merry-go-round. In the real atmosphere, airspeeds are generally much less than the angular speed of points on Earth, although perhaps not near the poles. So, the air goes round in circles, so to speak, creating circular weather systems.

Thus, the Coriolis force can set the winds in circular loops, such as that seen on surface weather maps in the form of rotating vortices, highs, and lows. These rotating vortices are a direct result of Earth's rotation, specifically of air trying to move in a straight line on a rotating Earth.

On Earth, the Coriolis force is actually a very weak one and causes significant displacement to the right (or left) only when the air has traveled a large distance. The circular motion of water going down a bathtub or sink drain has nothing to do with the Coriolis force, as the water can rotate in either clockwise or counterclockwise direction depending on the initial spin in the drain of the tub or sink. In fact, the circular motion is the result of centrifugal force, the force produced by the tendency for the water to fly outward—again, a kind of inertial force. Centrifugal force is also a factor in the atmosphere, because Earth is rotating about its axis. However, centrifugal force on Earth's surface is a very small one that amounts to a tiny correction to the gravitational force, although centrifugal force imposed by the Earth's rotation acts in the opposite direction to gravity.

Therefore, the Coriolis force is not evident in the atmosphere on small scales, although it operates on all scales, albeit a small effect in normal daily life. It may seem odd to imagine, but a baseball hit, say, toward center field experiences this Coriolis force toward its right and therefore undergoes a very small deflection toward right field by a measurable but unnoticeable amount, just a few centimeters (an inch or two). For ballistics or rocketry, a calculation for Coriolis deflection of the object is a necessity for accurate navigation. Had the baseball been hit so hard it traveled from Pittsburgh to Philadelphia, it would have undergone a sizable displacement resulting from the Coriolis force, perhaps many kilometers. Because this force is so weak, one would not expect to see any effect of the Coriolis force within a bucket of water or even a swimming pool, although some entertainers may attempt to erroneously demonstrate this effect on that small a scale.

Looking down from space at Earth from above the North Pole, we see Earth rotating counterclockwise, as in the merry-go-round illustration (Figure 1.3). However, looking down from space above the South Pole, we see Earth rotating clockwise. This change-in-rotation sense when looking down over the North Pole versus looking up from under the South Pole is responsible for the Coriolis force acting in opposite directions in the two hemispheres.

Earth is a sphere, not a merry-go-round

After his exciting day on the merry-go-round, Jerry had a vivid dream that night. He was in a museum with his mother in a large room in which a large model of the rotating Earth filled the center of the room. Latitude circles, continents, and political boundaries were drawn on the surface of the globe, allowing Jerry to locate the United States. Jerry and his mother stood on a balcony that surrounded the globe on all sides, allowing a view from above the globe, whereas visitors below at floor level could only view the globe from the side. Jerry leaned over the balcony looking at the top of the globe: the North Pole. He wondered why the Northern Hemisphere was always depicted on the top of the globe because it seemed to him, correctly, that the South Pole could just as easily have been at the top.

As he continued to look down on the globe, Jerry marveled that the top surface of the rotating sphere reminded him exactly of the merry-go-round, which was also rotating around a central axis. When his mother's back was turned, Jerry climbed down on top of the globe and situated himself close to the pole. Holding his arms outstretched, Jerry turned on his own axis. This reminded him of the game he used to play in which children would whirl around and round until they became so dizzy that, laughing, they fell to the ground. Here, the globe was making only about one revolution each minute, so the effect of this rotation was unable to make Jerry dizzy. Jerry noticed, however, that the seemingly flat top of the globe was somewhat of an illusion, as Earth's surface curved away from the pole and downward toward the floor. Intrigued, Jerry moved down the globe, like a fly on the wall, finally situating himself on the equator. Jerry continued to go around with Earth as it turned, but he was no longer turning on his axis. It seemed to him it was just like driving in a car along a straight road over a hill rather than rotating on his axis.

Jerry noticed that the top surface of the globe resembled the rotating merry-go-round. However, while looking down from the balcony, he also noticed that because of Earth's spherical shape objects near the equator did not face upward toward the viewer on the balcony but toward the side. Imagine Jerry, as in Figure 1.4, standing exactly at the North Pole. As Earth turns he twirls on his axis like a spinning top. The result of Earth's perceived sphericity is that while near the pole he experiences a rotation around his own axis; he does not experience any such rotation at the equator but merely follows Earth as it rotates. If we take a globe, the kind that some people or libraries like to keep, and paint latitude circles, we would not see the latitude lines near the equator as we looked down from above either pole. The surface of Earth near the top part of the globe spins around its axis like the merry-go-round but looks less and less like a flat disk rotating around its axis as one shifts his gaze away from the pole and looks at the lower latitudes. What this means is the Coriolis force, unlike that for a flat merry-go-round, is a maximum at both poles but then decreases toward the equator from both poles and vanishes at the equator, which is at 0° latitude.

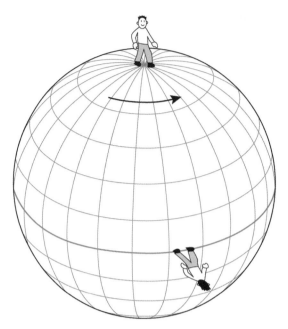

FIGURE 1.4. Jerry standing at the North Pole of the model globe, rotating counterclockwise, and at the equator.

Not surprisingly, rotating systems such as high and low pressure systems and hurricanes are feeble or nonexistent near the equator. A baseball hit in Bogota, Colombia, which is near the equator, would experience much less displacement due to the Coriolis force than a ball hit in Pittsburgh.

Buoyancy, the antigravity force

Most everyone realizes that warm air rises and cold air sinks. Attics tend to be warmer than basements. More precisely, air that is less dense than its surroundings rises and air that is denser than its surroundings sinks. In a broad sense this also applies to the atmosphere.

We can see this quite clearly by looking at the way a fireplace in a living room affects the surrounding air. When the fire burns, the heated air becomes less dense than its surroundings, the density of the column of air in the chimney is decreased, its weight is decreased (as is the pressure), and it rises through the chimney as the air from the surroundings is pushed toward the fire by the pressure difference between the lower pressure in the fireplace and the higher pressure outside of it. Thus, as air rises through the chimney, air near the floor outside the fireplace is seemingly sucked into the base of the fire by the pressure gradient force, resulting in a flow of air from the room into the fireplace and then up the chimney. Because of the pressure drop, the air experiences a force imbalance in which the air from the room, initially at rest, is pushed into the fire with an attendant reduction in density as it is heated by the fire. The air that flows into the fire from the room can be said to converge inside the fireplace. Above the chimney, the heated air spreads out into the environment; it can be said to be diverging.

Air rises in the chimney because with the reduced air density the upward pressure force exceeds the downward force of gravity on the heated air; this effect is referred to as convection, which is fueled by buoyancy or positive buoyancy in this case because the heated air is less dense than its surroundings and the air rises. The net force difference between the upward pressure force and the downward gravity force is referred to as the buoyant force. As will be discussed later, positive buoyancy (buoyant force directed upward) is responsible for the growth of cumulus clouds. Negative buoyancy (the air is more dense [colder] than its surroundings) is also possible in the atmosphere, as in the case of mammatus clouds (discussed in Chapter 3), which often accompany thunderstorms and are essentially upside-down cumulus clouds. Buoyancy in regard to atmospheric stability will also be discussed.

The key word here is *surroundings*. Colder air is denser than warmer air at the same pressure, which is to say at the same level in the atmosphere. Air at the level of a jet plane may have a temperature as low as –70°C, but it is still less dense than warm tropical air at the surface. If we were to lift some of that tropical air from the surface to jet plane level, it would cool so rapidly by expansion that it would likely become denser (and colder) than its surroundings and therefore fall back toward Earth. It would therefore still be negatively buoyant.

Convergence, the atmosphere's pump

Expanding on the idea of the fireplace, let's imagine a bonfire outside, perhaps made by Jerry's Boy Scout troop. As in our living room with its fireplace, the heated air in the fire (marked by its smoke) will rise and be replaced by converging air from all sides of the fire, as in the left side of Figure 1.5. Watching the smoke rise, Jerry sees that the heated air starts to spread out well above the fire. This is also true of the larger-scale weather patterns; as air converges below, it must diverge above.

Now, let's imagine that Jerry's bonfire is much larger, affecting an area the size of Pennsylvania and Ohio, so that the Coriolis force now becomes important on that scale, requiring the air to undergo a displacement from an apparently straight path. The result is that the air flowing toward the base of the fire from its surroundings is displaced to the right by the Coriolis force (in the Northern Hemisphere), as shown by the curved arrows on the right

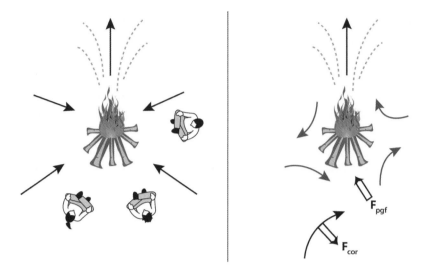

FIGURE 1.5. Three people sit around a bonfire (left side). Heated air from the fire rises vertically (heavy arrow) while air from the surroundings converges into the fire (horizontal arrows). On the right side is a bonfire hundreds of miles across, similarly causing a pressure gradient force directed toward the fire (double-shafted arrow). The heated air flows toward the fire, converges, and rises (heavy solid arrow), but on this larger scale, it also experiences a significant Coriolis deflection to the right (curved arrows), resulting in the initiation of a counterclockwise circulation around the fire. In this figure the pressure gradient force (F_{pgf}, the double-shafted arrow directed inward toward the fire), and the Coriolis force (F_{cor}, illustrated by the separate double-shafted arrow below the figure and directed to the right of the air motion), are labeled accordingly.

side of Figure 1.5, imparting to the air a circular, counterclockwise motion rather than a radial one as is illustrated in the left side of this figure. What is then being created is a counterclockwise swirl or vortex. This is called cyclonic motion.

Pressure versus Coriolis forces: the uneasy standoff

Note, however, as the air is displaced more and more to the right of its original direction, inward toward the fire and toward lower pressure, the Coriolis force, which acts to the right of the motion (at right angles to the direction of the wind and proportional to the speed of the air), causes the path of the air to bend increasingly toward its right. The result is that the air accelerates in the direction of the lower pressure, while the Coriolis force, proportional to the wind speed, pushes the air increasingly toward its right and toward a direction increasingly parallel to the isobars. As the air starts to move along (rather than across) the isobars toward lower pressure, the turning of the winds causes the Coriolis force to act increasingly in the opposite direction of the pressure gradient force, as illustrated on the right-hand side of Figure 1.5. The result of the turning of the winds is that the Coriolis force begins to act increasingly in the opposite direction of the pressure gradient force. In so doing, both the pressure force, which acts toward the lower pressure, and the Coriolis force, which is toward the right of the motion but becomes increasingly directed toward higher pressure, start to oppose each other until a close balance is achieved between the Coriolis and the pressure forces, each acting in opposite directions. Any change in the pressure gradient force requires the winds to move across isobars and therefore for the Coriolis force to change in order to accommodate the new balance. In a sense this balance resembles the virtually opposing forces depicted in the tug-of-war illustration (Figure 1.1b).

Of course, pressure falls in the atmosphere are not generally produced by bonfires. We can now generalize the concept in the following way. Imagine that for some reason surface pressure decreases at some location, thereby causing the pressure gradient between the area of pressure falls and the surroundings to increase. All other things being equal, the pressure gradient forces must temporarily exceed the Coriolis forces, thereby requiring the air to begin accelerating into the area of lower pressure, as illustrated on the left-hand side of Figure 1.5. In turn, the air becomes deflected to the right into a clockwise rotation as on the right-hand side of Figure 1.5. This process continues until its increased speed allows the Coriolis force to once again approach an approximate balance with the pressure gradient force—at which time the air direction has become parallel to the isobars. An analogous situation occurs if the pressure rises, except that the air motion is away from the pressure rises, ascends, and turns clockwise instead of counterclockwise in response to the Coriolis force. Since this continuous adjustment is constantly occurring, one is not normally aware of the aforementioned processes when viewing a weather map.

So, let us summarize the important point being made here: *At the surface, the air will tend to converge, experience increased cyclonic rotation and ascend in regions of falling surface pressures and diverge, experience increased anticyclonic motion, and descend in regions of rising surface pressures.* This attempt by the atmosphere to restore equilibrium is responsible for the creation and movement of "highs" and "lows."

What causes this continual imbalance? The air is always moving and, in so doing, it causes temperature changes everywhere, most importantly where the temperature gradients are large. In turn, these temperature changes, in affecting the weight of the air column, require the pressures to change at each point where warmer or colder air is being moved in. The process by which this occurs will be further discussed below and in Chapter 2.

The geostrophic wind speed is the wind speed that satisfies the mathematical balance between these two major forces, Coriolis and pressure gradient forces. Geostrophic balance implies that the wind blows parallel to the isobars with higher pressure to the right of the motion in the Northern Hemisphere and to the left of the motion in the Southern Hemisphere. The idea that higher pressure can be found to the right of the air motion was a very important principle to sailors during the 19th century, becoming a reliable guide to avoid storms at sea. Formulated in 1857, this principle is known as Buys Ballot's law, named after a Dutch meteorologist.

The larger the pressure gradient, the larger the Coriolis force must become to balance the pressure gradient force; thus, the closer the spacing of isobars (lines of constant pressure on a weather map), and the higher the wind speed. In geostrophic balance, the isobar spacing is inversely proportional to the wind speed, so that we would expect that high wind speeds would be encountered in regions (say around a low pressure system) where many isobars are spaced close together and weak winds where the isobars are few and far apart. Except in the lowest few hundred meters near the surface, where friction imparts a significant drag on the air speed, the geostrophic wind speed is an excellent approximation to the actual wind.

As mentioned, the Coriolis displacement (but not the Coriolis force itself) is negligible on the scale of the swirl of water going down the drain in a sink or bathtub. Indeed, this force is unimportant for dust devils or tornadoes. In those types of small-scale vortices, the pressure force is balanced by the centrifugal force. In rotating systems, the latter force acts outward and the pressure force acts inward toward the center of the vortex. Unlike geostrophic balance, however, the pressure at the center of the vortex is always lower than outside the vortex in centrifugal balance, although the rotation can be either clockwise or counterclockwise. Clockwise-rotating (anticyclonic) tornadoes do occur, although they are relatively rare. Balance between pressure and centrifugal force is called cyclostrophic balance. Centrifugal force even plays a role in intense smaller-scale cyclones such as hurricanes. However, we will not be concerned with this type of balance in subsequent discussions.

Friction: the atmosphere drags its foot

As it moves, air rubs against the surface of the earth, causing a drag on the wind near the ground. Not surprisingly, this drag slows down the air. Indirectly, frictional drag is responsible for the fact that wind speeds are lower near the surface than at high elevations, although the drag effect acts directly on the air only over the lowest few hundred meters. In slowing down the air, the Coriolis force, which is proportional to the wind speed, is also decreased by the drag, allowing the pressure gradient force to dominate, thereby pushing the air toward lower pressure and away from higher pressure; that is, in a direction down the pressure gradient. As in the example of the tug of war (Figure 1.1b), the atmosphere quickly adjusts to this force imbalance so that the frictional force aids the Coriolis force in balancing the pressure gradient force. In so doing, a new and somewhat shaky equilibrium is established between the three forces—Coriolis, pressure gradient, and friction—yet with the wind direction aimed slightly toward lower pressure rather than being exactly parallel to the isobars. This effect is only to be noted near the ground; at altitudes several hundred meters or more above the earth's surface, the winds remain in approximate geostrophic balance. Frictional force is also strongest during the day when the air that is being slowed down near the surface gets mixed with faster-moving air farther aloft as the result of surface heating. The surface wind speed typically will be about 1/3 of the geostrophic value over land and 2/3 of its value over a large body of water during the day, while approaching calm at night over land.

One can see the effect of this frictional drag on a surface weather map where the winds are blowing at a small angle across the isobars toward lower pressure (instead of parallel to the isobars), thereby causing a convergence of this surface air in the vicinity of low pressure centers and a divergence in the vicinity of high pressure centers. Although this shallow convergence is not of major importance in producing precipitation at middle latitudes, it can have some important effects on local weather, as will be discussed in subsequent chapters. The friction layer is shown schematically in Figure 1.6.

Highs and lows: anticyclones and cyclones

In the Northern Hemisphere, counterclockwise motion is referred to as cyclonic and clockwise motion as anticyclonic; in the Southern Hemisphere, where the Coriolis force acts to the left of the motion, clockwise (counterclockwise) motion occurs around a low (high) pressure center. Lows in the Southern Hemisphere are still referred to as cyclones, and highs are referred to as anticyclones, just as in the Northern Hemisphere even though they rotate in the opposite sense from lows and highs in the Northern Hemisphere.

Let us return to the counterclockwise flow in Figure 1.5. Air in geostrophic balance flows around the low center with higher pressure on the right, moving toward neither higher nor lower pressure, requiring air to move parallel to the isobars. Accordingly, we could imagine a series of somewhat concentric contours, each labeled with the value of pressure, or isobars, with air blowing approximately parallel to the isobars with high pressure on the right in the Northern Hemisphere.

The global picture

While lows and highs are usually not created by heating, (although heated surfaces of deserts do tend to develop low pressure at the surface referred to as heat lows), weather systems are created by the latitudinal imbalance of heating and its accompanying temperature gradients, resulting from more sunlight reaching lower rather than higher latitudes. By this we mean that the sun heats the earth more at lower latitudes than it does at the higher latitudes. In turn, the earth heats the air above it, typically up to a kilometer or two (Figure 1.6), above which the heat is circulated by air motions. It is somewhat surprising to think that our atmosphere, although indirectly heated by the sun, is more directly heated by the underlying surface of the earth.

This difference in heating between lower and higher latitudes creates temperature gradients, generally with decreasing temperature from lower to higher latitudes. Such temperature gradients in turn create pressure gradients that cause warm air to move generally from warmer to colder regions and cold air to move from colder to warmer regions. The unequal amount of sunlight received, more in the tropics than at high latitudes, creates not only the temperature difference between low and high latitudes but also pressure differences that, as we have seen in the bonfire analogy, are the driving force for atmospheric motions. The unequal heating and its attendant pressure distributions cause exchanges of warm and cold air across the latitudes in which warm air moves poleward and cold air equatorward, thereby preventing the tropics from heating up or the polar regions from cooling down indefinitely. Similar motions cause exchanges of heat between lower and higher latitudes in the oceans, acting through ocean currents. (The Gulf Stream, which moves from the U.S. Gulf of Mexico through the Florida Straits northward along the East Coast of North America toward Greenland, is one such mechanism for conveying warm water to cooler regions.)

Because both the temperature gradients and the Coriolis force are much weaker at low latitudes than at higher latitudes, the large-scale motions engender weak vortices (except for hurricanes) at these low latitudes. Instead, the overall tropical weather pattern is dominated by a gentle rising and sinking, whereby air rises in the equatorial regions (and at low latitudes over the land masses in summer) and sinks about 30° of latitude away from the equator. This type of circulation is called the Hadley cell after its discoverer, a 17th-century British astronomer. Because we are going to concentrate on midlatitudes (to be discussed below), particularly on the weather in the northeastern United States, we will leave the subject of global circulations.

Between about 30° and about 60° latitude is a region known as midlatitudes. In this latitude band, the Coriolis force and the temperature gradients are much stronger than at low latitudes (although the latter may be broken into unusual patterns over the land). As a result, the motions required to move warm air poleward and cold air equatorward at midlatitudes are much more chaotic than the aforementioned Hadley cell, whereby the patterns are characterized by transient waves and vortices (highs and lows). As discussed throughout this book, the wind flow at upper levels is characterized less by vortices that move generally from west to east, as at the surface, and more by undulations (wavy-type patterns), which also move from west to east in tandem with the surface pattern. These vortices at low levels and undulations at upper levels are accompanied by spasmodic gulps of rising warm air and sinking cold air. In other words, the latitudinal exchanges of heat must be accomplished by cyclones and anticyclones near the surface and by undulations in the westerly winds at upper levels. We will concern ourselves with weather systems that affect midlatitudes, as will be seen in Chapter 2.

All this sloshing and surging of differing airstreams, moving poleward, equatorward, eastward, westward, and up and down, takes place over an atmospheric depth of roughly 12–15 km, a distance not much greater than that covered by the average jogger during her daily run. This layer over which this sloshing, our weather, occurs is called the troposphere. The troposphere is bounded rather sharply at the top by a somewhat porous feature called the tropopause (Figure 1.6), across which air is sporadically exchanged with the air above, the stratosphere. Looking up at a layer of cirrus clouds at 12-km altitude, one is struck by the fact that the air at that level—indeed the air at any level in the troposphere—was once at ground level, yet its visit to the surface may have occurred weeks earlier. All air in the troposphere is recycled from top to bottom and from bottom to top and in time scales measured in weeks, yet mostly adhering to geostrophic balance.

In midlatitudes, the latitudinal temperature gradient causes warm air to move generally poleward and ascend from the lower troposphere to the upper troposphere. In so doing these rising airstreams are subject to Coriolis deflection to the right (in the Northern Hemisphere); that is, toward the east, resulting in the air gaining a rather strong westerly component, sometimes

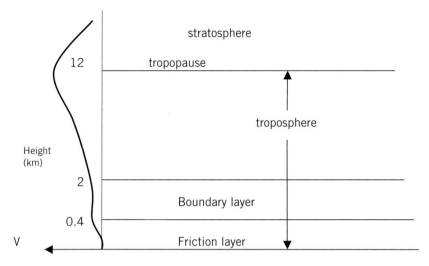

FIGURE 1.6. Schematic structure of the lower atmosphere, showing the various layers referred to in the text with typical elevations on the left. At the far left is a typical wind speed profile versus height, with increasing speed toward the left (indicated by the letter V).

reaching 100 m/s (200 kt). Another way to look at this is that Earth and its atmosphere are rotating much faster at the equator, about 22,000 km (Earth's circumference) per day, than at midlatitudes where the circumference of Earth along those latitudes is much shorter. Air moving at these speeds in absolute space may arrive at midlatitudes, where Earth's surface is moving more slowly around its axis, as can be seen from the shorter lengths of the latitude circles in Figure 1.4. Hence, the initially high momentum of this transported air tends to be conserved, and therefore it moves much faster from west to east relative to the motion of Earth when it reaches midlatitudes. This description is completely consistent with the Coriolis force; it's just another way of describing it. Yet this explanation does make obvious the explanation for fast-moving jet stream winds at high levels in midlatitudes.

The strongest winds at jet level typically reside just beneath the troposphere (Figure 1.6). This rapidly moving river of air, typically only a few hundred kilometers in breadth, is largely confined to the stormy midlatitude band between about 30° and 60° from the equator, and is called in popular language the jet stream. An intimate connection exists between the wind speed (specifically, the vertical variation of the wind speed with height, which is largest just below and just above the maximum winds within the jet stream) and the horizontal temperature gradient. For that reason, colder air is found on the left side of the jet (usually the poleward side) and warmer air is found on the right side (usually the equatorward side), at least below the top of the troposphere.

The opposite situation, in which air descends from upper levels and moves toward lower latitudes, is likewise subject to deflection to the right (toward the west), but the flow at low levels is slowed down by friction and by mixing up from below the slower-moving air at low levels with the faster-moving westerly air stream aloft. The result is that the low-level flow still maintains a weak westerly component, because of the mixing down of the faster-moving air. Consequently, even near the surface the prevailing winds are from the west (even in the Southern Hemisphere), albeit much weaker than at jet stream level, typically 5–10 m/s (10–20 kt).

Unlike cyclonic and anticyclonic vortices characteristic of the surface weather map, the northward and southward surges of air at upper levels create undulations (or waves) in the westerly flow, which move eastward approximately at the average westerly speed component about halfway between jet stream level and the surface. These waves in the westerly current correspond to the highs and lows in the lower troposphere, as the upper- and lower-level weather systems are strongly coupled, whereby the lows at jet stream level correspond to the troughs (equatorward extensions of the westerly current) and highs to ridges (poleward extensions of the westerly current).

Why it rains or snows

As air rises, it moves from higher to lower density (recall the analogy with the pillows) and in so doing it expands and cools. The reverse is true for air descending: It compresses and warms. This process is similar to that occurring in a refrigerator or air conditioner wherein expansion and compression alternately cool and warm a gas called a refrigerant: the former process cooling the room or refrigerator interior and the latter being expelled as waste heat. Were we to lift a parcel of air from the surface without condensation, it would cool at a rate of approximately 10°C per kilometer. As mentioned, air lifted to the tropopause would become very cold indeed, something like −70°C, typical of temperatures at that altitude.

In ascending, though, the cooling air can reach condensation, at which point droplets form. Condensation releases heat (called the latent heat of condensation, the opposite of evaporation in which liquid water absorbs that latent heat). In a cloud, condensation produces many small droplets, initially too light to fall very far before they evaporate. Condensation creates the film of droplets on the outside of a glass of iced tea on a summer's day, or on the inside of windows in a humid shower room. Cloud droplets can grow large enough to fall to the ground as precipitation, provided that they grow to a certain size and become heavy enough to fall. We might notice that some drops on the outside of our iced tea glass succeed in rolling down to the table surface after they combine with other drops, thereby forming a heavier drop capable of descending the length of the glass.

Condensation takes place when the temperature decreases to the temperature of saturation, called the dewpoint, which is a measure of the actual water content of the air. A crude measure of the relative humidity can be assessed as the difference between the actual temperature and the dewpoint temperature; the larger the value, the drier the air. More precisely, relative humidity is approximately the ratio of the actual water vapor content of the air divided by the saturation water vapor content, which is the maximum possible water vapor content of the air at that temperature. A zero difference between temperature and dewpoint corresponds to 100% relative humidity with respect to water saturation. Droplets can form in the atmosphere on small dust particles when the relative humidity is high but less than 100%. This is often the cause of haze on a humid day.

In clouds, this coalescence occurs if there is a sufficient density of droplets that can reside in the cloud long enough to form larger drops, primarily through the coalescence mechanism—small droplets colliding with each other or with larger drops. This process of drop formation sufficient to produce rain requires not only a high density of drops but, as stated above, also some residence time in the cloud to allow the drops to become large enough to fall out. Accordingly, we often see clouds in the sky but experience no precipitation from them. Fair-weather cumulus clouds, for example, generally do not last more than a few minutes before they begin to dissipate, so naturally we would not expect precipitation from such clouds. In most clouds the droplets are so small that once they descend below cloud base they immediately evaporate.

Where does the rain come from?

While it is technically slightly imprecise to say so, it is effectively true that the warmer the air, the higher amounts of water vapor can be stored within it. Large amounts of water vapor in the tropical atmosphere are responsible for the heavy rainfall often experienced at lower latitudes or even in summer from thunderstorms at higher latitudes, whereas areas such as Greenland and the Antarctic are essentially deserts, albeit covered with snow and ice. Accordingly, most water vapor is stored in the lowest 2 or 3 km, the warmest part of the atmosphere, and therefore most of the precipitation that falls to the ground originates within these lowest layers.

Although relatively little moisture resides above the first few kilometers, it is clear from looking at clouds (Chapter 3) that at least some water can exist at much higher altitudes, even at very high altitudes in the form of ice crystals. Precipitation is therefore a byproduct of the rising airstreams, referred to above.

Clouds themselves do not hold most of the water that falls to the ground, but simply act as conduits through which rising air produces precipitation in a continuous process. Water is supplied to the cloud from below through the process of low-level convergence of the air, as previously discussed. The cloud, in other words, is similar to the faucet in a sink as a conveyor of moisture from a source, but not a basic source itself. In the atmosphere, that source is moisture in the air near the surface, and the water is pumped up by convergence from those lowest layers. The cycle of evaporation from the earth and precipitation from clouds occurs with about as much water going up as vapor as is falling from precipitation, when averaged over the entire surface of the earth. As such, the residence time of this water in the atmosphere is about one week.

Atmospheric stability and instability: convection

Returning for the moment to Jerry's bonfire (Figure 1.5), we see the effect of buoyancy operating to lift the heated air. Being less dense (and warmer) than the surroundings, the heated air experiences an upward pressure force that somewhat exceeds the downward force of gravity, leading to an acceleration upward as long as the buoyant force (the difference between the upward pressure force and the downward force of gravity) remains positive (upward). The process in which heated, less dense air is lifted by the buoyant force is called convection.

Another good example of convection, in addition to the fireplace, is a cumulus cloud. We have discussed why this colder air at high altitudes does not sink. A parcel of air lifted to some higher level generally may remain colder than the surroundings, so that it would be negatively buoyant and therefore, upon release, sink downward. Thus the atmosphere in this respect can be said to be stable enough to endure up and down motions. By stable atmosphere we mean one in which a parcel of air, initially in equilibrium but subsequently lifted to find itself colder (more dense) than its surroundings, sinks downward until it reaches the point where it started, in an equilibrium state whereby the pocket of air again has the same density as the surroundings. The same is true for a pocket of air pushed downward, where in stable conditions it would usually find itself warmer (and therefore less dense) than its surroundings and thus experience an upward buoyancy force, pushing the air back to its original level where it would be once more in equilibrium with its surroundings. (It is worth emphasizing here that to say colder air is denser than its surroundings and warmer air less dense than its surroundings is only true when the air pockets are evaluated at the same level as their surroundings.)

Ascent or descent due to differences in density between the inside of a pocket of air and its surroundings can be referred to as convective instability. Convection is simply the vertical (up or down) motion of air resulting from buoyant forces caused by density differences between a parcel of air and its surroundings. Convectively stable air, if lifted or lowered, will thus return to its original altitude, like pushing (or pulling) on a bed spring. Unstable pockets of air will continue to rise or fall until they finally do find equilibrium with the surroundings. Most of the atmosphere at midlatitudes is convectively stable. However, the atmosphere may undergo local convective instability, of up and down motion, whereby the lifted air finds itself warmer (less dense) than the surroundings and would continue to ascend. This type of process, constituting a convectively unstable situation, drives cumulus clouds to grow to the point where they may cause local thunderstorms. Negative buoyancy may even drive cumulus clouds downward when they are colder than their surroundings. We will illustrate this peculiar situation of upside-down cumulus clouds in Chapter 3.

Temperature generally decreases with height up to the tropopause, which acts as a somewhat porous cap to the transient weather below. Here and there and over shallow layers the temperature may actually increase with height, most notably at night near the ground. These layers where temperature increases with height are called temperature inversions. Inversions are usually shallow but highly stable air layers, in the sense that a parcel of air lifted into an inversion will invariably find itself colder than the surroundings and subsequently sink back to its original level.

On the large scale, occupied by highs and lows, it is not appropriate to think in terms of a parcel of air being lifted or lowered, as the mass in that case might extend over several thousands of kilometers in the horizontal and several kilometers in the vertical. Nevertheless, as we will see in the next section, even large-scale masses of air, which ascend and descend, are responding to a kind of buoyant process.

Since air tends to rise faster once condensation is achieved (due to the release of latent heat to the air), it follows from continuity that if air in cloudy, precipitating areas is rising faster than it descends in clear areas, the areas of ascent and precipitation must be smaller than the areas of more gentle

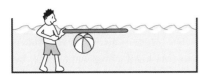

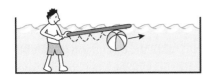

FIGURE 1.7. (top) Jerry releases a soccer ball underneath the water and watches it rise rapidly due to positive buoyancy. It surges through the top of the water into the air, narrowly missing his face. (middle) Jerry holds a surfboard horizontally over the submerged soccer ball, thereby preventing its ascent. (bottom) Jerry tilts the surfboard slightly upward toward the right, so that it makes a very small angle with the horizontal. The ball moves sideways toward the higher end of the surfboard, still responding to buoyant forces but restricted from moving vertically.

descent. Cloudy areas, those associated with precipitation, thus tend to be much smaller than cloud-free areas.

In general, instability involves an imbalance of forces whereby a net force moves the air up or down or sideways. Generally, the atmosphere is stable with respect to vertical motions (air rising straight up in convection), except over small areas (meters to tens of kilometers in size) in which these convective (cumulus) clouds are embedded.

Figure 1.7 (top) illustrates convection, ascent due to buoyancy. Here Jerry releases a soccer ball that he has placed under the surface of the water in his neighborhood swimming pool. The figure shows what happens when Jerry releases the ball from some distance under the water surface. Since the interior of the soccer ball is air, it is much less dense than the water surrounding it even when considering the heavier outer covering of the ball, and there is a net force imbalance between the pressure force pushing the ball upward and gravity holding it down. The ball moves rapidly upward, accelerating as it rises, and pushes through the surface with a splash, spewing water behind and above it. Ascent can be quite violent, as Jerry might discover if he keeps his head in a position above the rising soccer ball. As the ball surges through the surface of the water, it finds itself surrounded by air, a much less dense medium than water or the soccer ball. Accordingly, the ball experiences a downward buoyant force but because objects in motion tend to remain in motion, the ball continues its upward trajectory as gravity, now the dominant force on the ball, slows its upward motion and then returns it to the surface. This is an illustration of convective instability, although in the atmosphere the convection is not taking place through the ascent of soccer balls but through buoyant blobs of lighter, warmer air (convection).

Atmospheric stability and instability: slantwise ascent

While this illustration of vertical convection evokes images of warm, moist thermals rising over a heated land surface, the larger-scale weather systems, highs and lows, do not respond directly to this kind of buoyant force. Yet even on a global scale cold air sinks and warm air rises, which is nevertheless a kind of buoyant instability. As an analogy, let's imagine that Jerry places his surfboard directly over the ball when it is submerged, while keeping the surfboard absolutely horizontal. The ball, of course, will lodge itself just be-low the surfboard, as in Figure 1.7 (middle panel). The surfboard serves as a kind of stable lid, inhibiting the ball's ascent; recall the discussion of stability for up and down motions. Yet, the ball is still experiencing a net upward force due to buoyancy, which cannot be translated into motion of the ball because the material surface of the surfboard is pushing back against it.

Now imagine that Jerry tilts the surfboard upward just a bit, say at a slight angle from the horizontal, as shown in Figure 1.7 (bottom panel). (For the purposes of later discussion, imagine a tilt of only about 1%.) In this case, the ball will skitter along the bottom of the board moving toward the higher end of the board. The ball responds to buoyant forces, but because of the stable lid it is forced to move sideways. So also in the atmosphere the air is mostly stable with respect to vertical ascent (air moving straight upward as in a cumulus cloud), but it exists as a controlled instability within a controlled, unstable environment with respect to sideways (nearly horizontal) ascent. Warm air still rises and cold air descends, but in a more sideways fashion rather than straight upward or downward, requiring thousands of kilometers of sideways movement and perhaps days to ascend from near the surface to the upper part of the troposphere. The overall process resembles the tilted surfboard example, whereby the air, in moving from warmer to colder temperatures, tends to rise along imaginary sloping surfaces, and air moving from colder to warmer temperatures tends to descend in the same manner. The tilt, the sloping streams of ascent and descent, is most prominent in regions of strong temperature gradients. The angle of tilt of the surfboard is analogous to the magnitude of the horizontal temperature gradient in the atmosphere.

So, air streams at midlatitudes mostly rise or descend in a slanted mode, sometimes called slantwise convection. This horizontal type of motion is a kind of sideways instability, which meteorologists call baroclinic instability, though we need not adopt this fancy term. Instead, let's just refer to this instability as sideways instability, in contrast with convection, which is really a vertical instability. We adopt this term, sideways instability, to differentiate it from the instability that allows cumulus clouds to grow by convection (vertical instability), which is a more localized process. In subsequent discussion we will refer to vertical motion in the context of sideways ascent as the vertical component of that sideways motion, whereas in talking about buoyant convection, the vertical motion is understood to be directly upward rather than on a slant path.

Convective instability versus slantwise instability

Both sideways and convective instability can be occurring at the same time. Say Jerry left his beloved surfboard on the driveway aside his house. That evening, his father accidentally rolled over the surfboard with his van, cracking it in several places. Jerry was heartbroken when he saw the damage; he was able to put two fingers through the hole in the middle of the board. Next day, when he went to the swimming pool and played with the soccer ball, watching it roll along the base of the board when he tilted it upward, he noticed that a few of the acorns that had been floating on the surface became trapped under the surfboard. As the ball rolled sideways, Jerry saw that one or two of the acorns rose vertically through the crack in the surfboard, thereby demonstrating the fact that both types of instability, sideways and convective, can exist simultaneously.

So it is also in the atmosphere. Like Jerry's surfboard, the stable atmosphere can be simultaneously unstable with respect to larger-scale slant motions (the soccer ball in Figure 1.7) but can be porous (unstable with respect to convection) on a smaller scale (the acorns), allowing air to rise vertically in thermals, as in the top panel of Figure 1.7. The difference between slantwise instability and convective instability is scale. Convective instability is produced by buoyant force, whereby a positive (upward) buoyant force is exerted on a pocket of air that is less dense (essentially warmer) than the immediate surroundings at that level and negative (downward) buoyant force is exerted on a pocket of air that is more dense than its surroundings at that level. When a buoyant force is either positive or negative the pocket of air is said to be convectively unstable, and the air pocket will continue to rise or fall until some other force exerts a brake on its motion.

Convective updrafts in cumulus clouds are typically a few hundreds of meters in size. In thunderstorms, convective clouds are sometimes several kilometers in width, although they may grow to depths of 10–15 km or higher. While large cumulus clouds, such as those found in thunderstorms, can grow to such great depths, their horizontal and vertical extent are nevertheless small compared to the size of the highs and lows found on a weather map. Not all convection is associated with an upward force. As we will see in Chapter 3, negative buoyancy sometimes occurs in clouds, resulting in upside-down cumulus clouds.

In the surfboard illustration, the sideways (or slantwise) instability is produced by Jerry tilting the surfboard. In the atmosphere, however, this type of sideways instability depends on horizontal gradients in temperature, which create sloping density surfaces, just as in the tilted surfboard example in the lower panel of Figure 1.7. The strongest temperature gradients are found in the vicinity of the jet stream and surface cold and warm fronts. Here, the proverbial atmospheric surfboard is tilted upward toward the colder air at its steepest incline, but the surfboard is replaced by sloping density surfaces. These sloping density surfaces are accompanied by pressure gradients that move the air sideways, but also with an upward or downward component. Normally, this type of sideways instability is restrained in the atmosphere, although in some cases it can lead to the explosive development of cyclones (storms). We will expand on the idea of sideways instability leading to storms in the next chapter. We will also touch on the conditions for which the atmosphere becomes porous to convective instability.

The basis for sideways instability is that the movement of colder toward warmer air and warmer toward colder air acts to continually destabilize the atmosphere by changing the pressure field in the same way that the heated air in a fireplace changes the pressure gradient between the room and the fireplace. In turn, this change in the pressure field alters the relationship between the Coriolis and pressure gradient forces, thereby creating temporary imbalances between these two forces. Accordingly, the Coriolis force (and therefore the wind speed) is required to continually adjust itself to the changing pressure field, playing catch up. It must continually chase after the pressure gradient force. The bonfire analogy shows that in the Northern Hemisphere where the pressures are falling, air will be converging toward the area of falling pressures and therefore undergo a counterclockwise (cyclonic) rotation. Similarly, air diverges from regions where the pressure is rising, and undergoes a clockwise (anticyclonic) rotation.

Like convection, slantwise ascent may also involve air moving from the lower to the upper troposphere or from the upper to the lower troposphere,

FIGURE 1.8. Example of a cumulus cloud, which is fueled by positive buoyancy.

but this ascent or descent occurs with a slope upward or downward, typically much less than a 1% grade. This type of gradual slope ascent or descent is indirectly due to the fact that the atmosphere is very much shallower in the vertical than the horizontal dimensions of the highs and lows—tens versus thousands of kilometers. It is a sobering thought that our atmosphere is so very thin. If we were to wet a soccer ball by immersing it in water, the film of water clinging to the ball is, in relationship to the size of the ball, a fair analogy for the depth of the troposphere in relation to the size of Earth. The ratio of depth to width of a weather system is therefore less than 1%, so it is not surprising that the air rises at a rate typically much less than the speed of horizontal motions (centimeters per second versus meters or tens of meters per second for convective instability). The latter is manifested in the form of cumulus clouds and various types of dry convection, such as dust devils and invisible up- and downdrafts on which birds love to soar.

Two photographs illustrate the difference in these two types of atmospheric instabilities by their respective cloud forms, to be discussed in greater detail in Chapter 3. Cumulus clouds, typical agents of convective instability, are puffy and pillowlike, the ice cream castles in the air as described in the Joni Mitchell song, "Both Sides Now." An example of this kind of cloud is shown in Figure 1.8. Layer clouds, such as the stratiform deck of clouds shown in Figure 1.9, are typical of sideways instability and so are spatially

FIGURE 1.9. Example of a layer cloud, which is formed by sideways ascent.

extensive, resembling large pancakes, and reflect the very gradual ascent of the air within them, as in the soccer ball example in the lower panel of Figure 1.7. Aircraft flying through layer clouds are likely to have a much smoother ride than flying through all but the smallest cumulus clouds.

Pressure changes are important signals

Barometers found in private homes are often marked with dramatic labels—stormy, hurricane, fair, dry, and change—as a way of indicating what type of weather to expect with a particular range in pressures. It is certainly true that a barometer reading less than 29 inches of mercury (982 mb) would indicate or presage stormy weather, and a barometer reading of 30.50 inches of mercury (1032 mb) would likely accompany fair and dry weather. More important, however, are the current *changes* occurring in the pressure—whether it is rising or falling and how rapidly these changes are occurring. Conventional surface weather maps often use a symbol to indicate the near current surface pressure change (referred to by meteorologists as the pressure tendency), which is evaluated as a change over the previous three hours.

Where pressures are rising or falling, the pressure gradient must also be changing, and therefore the strength of the pressure gradient force as well. Where pressures are falling, the pressure gradient force change implies an imbalance between that force and the Coriolis force, such that the air near the surface is accelerated in the direction of the center of falling pressure, undergoing a component of motion across the isobars (rather than blowing parallel to them). As we have seen in the example of the large-scale bonfire in Figure 1.5, this convergent flow toward the center of falling pressures must involve a Coriolis deflection to the right and the creation of a counterclockwise (cyclonic) component in the flow to establish a new equilibrium between the forces. In a sense, areas of pressure falls are areas of nascent cyclonic development, low pressure formation, and future low pressure centers.

Likewise, pressure rises will accompany an imbalance between pressure gradient and Coriolis forces, requiring the air to accelerate away from the center of pressure rises, diverge, and experience a component of clockwise (anticyclonic) motion. Alternately stated, these areas of pressure rises will constitute areas of nascent anticyclonic development, high pressure formation, and future high pressure centers.

Convergent and divergent areas are therefore areas where the Coriolis and pressure gradient forces are somewhat out of balance produced by the changing pressure pattern. As stated above, these areas constitute seeds of future cyclonic and anticyclonic centers—lows and highs. This deviation from geostrophic balance is usually not very noticeable by any significant deviation of the wind direction from being parallel to the isobars, but the convergence and divergence patterns are reflected in the patterns of pressure rises and falls—the pressure tendency pattern. Because convergence and divergence are so intimately connected to the weather and to precipitation, these processes, as reflected in the pressure tendency pattern, are very important for understanding the weather. Not surprisingly, the pressure tendencies are likely to be largest where the sideways motion of the air is slanted upward at the steepest angle (as in the tilted surface board analogy of Figure 1.7), which is to say, where the air has the strongest vertical component. The accompanied sideways ascent is also likely to be the strongest in regions of strongest horizontal temperature gradients. Thus, pressure and temperature in this respect are intimately linked in the atmosphere. Surface pressure falls, upward vertical motion accompanied by cloud and precipitation, are intimately linked. Similarly, surface pressure rises and descending vertical motion correspond to the absence of cloud and precipitation.

Like convergence and divergence patterns, the vertical motion associated with the sideways (or slant) ascent or descent are not apparent or easily measured directly because they are so small compared to the horizontal wind speeds—just a few centimeters per second, about as fast as a speedy ant can walk. Indirectly, however, this ascent and descent are reflected in the pressure tendency pattern, which is easy to measure. The largest temperature gradients are to be found on the cold side of temperature fronts, which is where most of the weather action takes place. We might say, referring to Figure 1.7, that the atmospheric surfboard is tilted upward at the steepest angles (albeit small angles) on the cold side of temperature fronts, which typically lie below the jet stream. These ideas will be developed further in the next chapter.

An overview of cloud types and altitudes

During the latter part of the 18th and the early 19th centuries, European society was awash in the excitement over scientific and technological discoveries and inventions. Unlike today, ordinary members of the public were keen not only to stay current with the daily advances in science but also to become involved in it. Professional and amateur organizations sprang up, especially in England, wherein citizens and specialists would convene on a regular basis to discuss the latest discoveries and to present papers by members. In 1802, a London druggist by the name of Luke Howard presented a talk outlining his systems for cloud classification. Until then, attempts to classify clouds spawned a welter of confusing and somewhat illogical cloud classification schemes. Howard's scheme, a masterpiece of simplicity and physical logic, immediately caught on, earning him international recognition in the naming of clouds. While later modified to include additional levels and types of clouds, Howard's classification has been preserved, almost as it was conceived, to the present day.

Although the official cloud chart published by the U.S. Weather Service contains about 100 different types of clouds, these clouds are mostly variants of just a few basic types, which are classified by altitude levels and texture.

Table 1.1 contains a simplification of the basic chart type. Three levels of clouds are shown: high, middle, and low. Three types of clouds, each representing different processes, are shown: cirriform (ice crystals), stratiform (flat), and cumuliform (Latin for *heaps*, puffy with convective updrafts), with the prefix or suffix *nimbus*, which is affixed only to low clouds and refers to clouds that yield precipitation.

These various types of clouds summarized in the table constitute the characters in the weather drama, each with its own origins, implied atmospheric processes, and significance. We will describe these clouds in much greater detail in Chapter 3.

Cloud types used to be reported routinely and plotted on weather maps. These days, however, most observations are made with the aid of automatic

TABLE 1.1. Cloud types, altitudes, and their variants with abbreviations. Abbreviations for the variant types are set in parentheses.

Cloud type	Approximate altitude range (in meters)	Variants
High clouds (prefix: *cirro-*; ice crystal clouds)	Above 6,000	Cirrus (ci) Cirrostratus (cs) Cirrocumulus (cc)
Middle clouds (prefix *alto-*)	3,000 to 6,000	Altocumulus (ac) Altostratus (as)
Low clouds	Below 3,000	Stratus (st) Cumulus (cu) Stratocumulus (sc) Cumulonimbus (cb) Nimbostratus (ns)

weather stations that can record and transmit temperature, humidity, pressure, and pressure change, but not cloud cover or cloud type. The present system for recording data on weather maps is far more efficient than previously implemented with the aid of live weather observers. Some automatic weather stations actually do measure cloud base heights. This is a natural outcome of the fact that weather prediction models that contain banks of differential equations can be much more easily handled with the aid of computers than by human beings and for a much longer range into the future. Models, however, do not specify cloud type, and automatic weather stations that transmit data to these models do not have the capability to recognize and code cloud types. Even so, such models yield better forecasts than can be made by humans. However, good weather forecasters can make use of computer model outputs and add their own skills and insights to provide better weather forecasts than just the computer results alone.

These 10 characters shown in Table 1.1 are the stars of Chapter 3 and will play major roles in the weather drama discussed in that chapter. Abbreviations later used in this book are listed in parentheses under the variant column. An abbreviation for a clear sky (no clouds), for example, is cl. We will introduce two other cloud forms, variants of cumulus clouds, in Chapter 3: towering cumulus clouds (tc) and cumulus congestus clouds (cg). Another useful cloud descriptor that will not be referred to in this book is the term fractus (or fracto- as an antecedent), such as in fracto cumulus or fracto stratus, the former referring to small torn-looking cumulus clouds, harbingers of fair weather, or stratus that is not at all uniform as in Figure 1.8, but has ragged bases with many internal, adjoining edges.

Where do we go from here?

The atmosphere is in a tenuous balance between two large forces, the Coriolis force and pressure gradient force. Imbalances are nevertheless necessary for cyclones and anticyclones to move, develop, and decay, and for the transfer of heat from lower to higher latitudes.

In the next chapter, we will look at the weather map with its isobars, isotherms (contours of equal temperature), and frontal systems as manifestations of the imbalance of forces. The features of the surface weather map, in particular, will be introduced, and the necessary conditions for cyclone behavior, especially rapid cyclone intensification, and the favored geographical locations for cyclogenesis will be discussed. Finally, we will describe how surface cyclone development is ultimately self-limiting by processes occurring throughout the troposphere.

CHAPTER 2
CLOUD AND WEATHER PATTERNS

A surface pressure map

Weather maps, especially surface weather maps, can convey the current weather patterns and weather information in an appealing and visual form, if properly understood. In this chapter we will discuss how highs and lows are formed, how they move, and what controls their size and intensity. In so doing, we will build up the various plotted components on a weather map and then provide some simple examples of weather as viewed primarily on conventional weather maps with emphasis on storm development and the classic cyclone model. Once this model is understood, the behavior of the weather patterns and of storms should allow us to integrate the cloud we observe, the pressure tendency, and wind direction within the context of the cyclone model; a discussion of the clouds will be presented in Chapter 3. The object of Chapters 1 and 2 is to help the reader understand how the sky and cloud observations discussed in Chapter 3 fit into the greater scope of the weather patterns and what they portend. We can enhance the knowledge gained from these first two chapters by observing the sky and making some easily measurable weather data, subjects for the last chapter. First, however, let us review a couple of ideas that require emphasis.

We have seen that air movement on a rotating sphere leads to circular weather patterns. More precisely, large-scale eddies, highs and lows, result from air moving on a rotating sphere in which a close balance between the two main forces governing air motions, pressure gradient and Coriolis, is maintained; this balance requires that the winds blow approximately parallel to the isobars, lines of constant pressure, with higher pressure on the right and lower pressure on the left (along the direction of the wind) in the Northern Hemisphere. (Because the Coriolis force acts to the left of the motion, the reverse is true in the Southern Hemisphere, where higher pressure is on the left and lower pressure is on the right.) Wind speeds are directly related to the spacing of the isobars—the closer the spacing, the faster the winds. As discussed in Chapter 1, imbalances between these forces constitute a controlled instability, technically termed baroclinic instability by meteorologists but for our purposes referred to as sideways (or slantwise) instability. Although self-limiting, sideways instability can occasionally manifest itself in quite spectacular fashion in the form of explosive cyclone development. The latter will receive much attention later in this chapter.

For those who like to hike, a map showing elevation contours is a familiar and often valuable asset. A weather map, such as shown in Figure 2.1, is also a kind of topographic chart, but with the contours representing pressure rather than elevation. High and low pressure systems are analogous to hills and valleys on topographic maps. In this figure, the little wind arrows show the direction of the wind (blowing in the direction from tail to head) and the relative strength. Each little segment attached to the tail of the wind barbs represents a speed of 10 knots (abbreviated as kt; or 5 m/s); two tail segments indicate a wind speed of 20 kt (10 m/s).

For clarity, the small cross-isobaric behavior of the wind barbs, which is produced by frictional drag (discussed in Chapters 1 and 4) on the atmosphere acting over the lowest few hundred meters, is not represented in this figure. Recall, however, that frictional drag will tend to turn the surface wind direction (typically by 10° or 20° in angle) toward lower pressure, thereby causing shallow low-level convergence in regions where the isobars are curved cyclonically (they enclose lower pressure) and divergence where the isobars are curved anticyclonically (they enclose higher pressure). We will return to this idea of frictional convergence, as opposed to convergence due to pressure changes, later in this chapter.

Figure 2.1 represents a somewhat generic depiction of a developing low pressure system with a high pressure system off to the west. We will refer to this generic pattern, the open wave, as one pertaining specifically to the low pressure system with warm and cold fronts, which meet to form an upside-down L-shaped configuration of fronts that join at the center of the low pressure area.

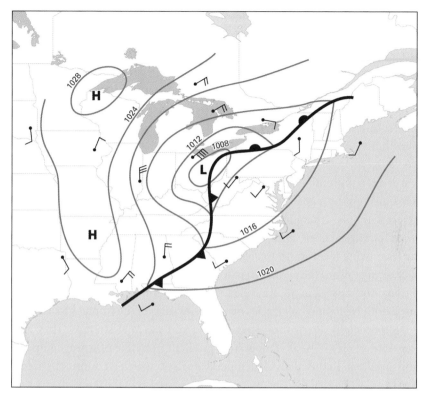

FIGURE 2.1. Schematic surface weather chart for the classic open wave cyclone, showing isobars (solid lines labeled in millibars), surface cold and warm fronts, and low and high pressure centers. Wind direction and speed are shown by the small flags: each tail segment equaling 5 m/s (10 kt) and the wind blows along a direction from the tail (flag) to the head. The cold front is marked by barbs, indicating the direction of the front's motion toward warmer air; the warm front is shown by half circles, indicating its direction of motion toward colder air.

Classic models such as this one have much use, but they will not exactly coincide with any actual pattern. No weather pattern, no matter how similar it may seem to another, is exactly identical. Although the open wave repeats itself, it does so with infinite variations. Once, over 50 years ago, it was thought that weather forecasts could be made by finding an exact match of the present pattern with one in the historical archives, the idea being that subsequent weather maps would correspond to those that evolved from the archived best fit. What was found, however, was that no pattern, however similar, fit exactly, so that the current pattern rapidly diverged from the archived best fit. Ultimately this method of analog forecasting proved essentially useless in the face of the chaotic nature of fluid flow.

Fronts: airstream boundary lines

Fronts are not arbitrary lines drawn on a map at the discretion of the analyst but are inherent in the physics of the atmosphere and in cyclone development. From the wind pattern in Figure 2.1, we can see that the wind flowing between centers of the high and the low pressure areas are embedded within southward- or northward-moving airstreams. Air flowing between the low center and the high pressure center to the west has a generally southward (northerly) component. Conversely, the air flowing east of the low pressure center is generally moving northward (a southerly airstream). These two rivers of air, representing airstreams with different origins (cold and warm), are separated on the map by boundaries, called fronts, which are drawn with different symbols. Triangular barbs show that the front is moving from cold to warm air (a cold front) and those with the half circles indicate that the boundary is moving from warm to cold (a warm front). In the former, the front is advancing toward the warm air; in the latter, the front is retreating poleward. Fronts not moving at all are called stationary fronts (not shown here) and have both barbs and half circles on opposite sides of the frontal boundary.

Superficially, the fronts separate cold and warm airstreams. They also separate an area of weak temperature gradient from that of strong temperature gradient. Rather than the temperature itself, which is continuous across the front, the actual discontinuity across the front is in the temperature gradient, which is always much stronger on the cold side of the front than on the warm side. The front is therefore not simply a boundary in temperature so much as a boundary in temperature gradient, as well as airstream origin. Later we will introduce another type of front called the occluded front.

This temperature pattern associated with Figure 2.1 is shown in Figure 2.2, which is identical to Figure 2.1 but with the wind barbs and pressure labels removed. Imagine the wind direction barbs, whose inclusion would clutter up the figure, as if the wind direction is parallel to the isobars with higher pressure on the right side of the motion and the wind speed inversely proportional to the spacing of the isobars in accordance with geostrophic balance between pressure gradient and Coriolis forces.

Here in Figure 2.2 west of the cold front, the air is moving from north to south and therefore from cold to warm, especially in the region just west of the cold front at the location marked with the letter C. Air blowing across isotherms from cold to warm is called cold advection. Cold advection is strongest in this example in the area marked by the letter C. Similarly, air blowing across isotherms from warm to cold is called warm advection, with the latter being strongest in Figure 2.2 just north of the warm front in the area

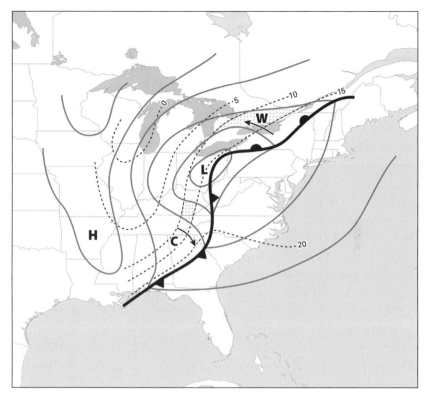

FIGURE 2.2. Same as Figure 2.1 without wind flags but with isotherms of surface temperature labeled in degrees Celsius. C and W symbols indicate, respectively, the locations of maximum cold and warm air advections. Arrows beside these symbols show the direction of the advection by the surface wind. Lightly shaded areas forming nearly parallel four-sided polygons near the letters W and C identify "advection boxes" referred to in the text.

marked with the letter W. Mathematically, the magnitude of the advection is the product of the wind speed times the gradient along the direction of the wind barb. Thus advections, warm or cold, are greatest where strongest winds (close spacing of the isobars) are moving across the strongest temperature gradients (isotherms close together). Aside from the cyclic temperature changes that occur in response to the daytime warming and nighttime cooling, cold or warm air advections are the primary factors influencing how the local surface temperature changes at a given location. Generally, we can assume that where cold-air advection is occurring, the air is becoming colder at that location and where warm air advection is occurring the air is getting warmer there.

A word about advection

It is essential to stress the importance for temperature advection, which is crucial for understanding the remainder of this chapter and, indeed, much of this book. Advection (whether temperature or any other quantity) depends on both the wind speed and the magnitude of the gradient, in this case the temperature gradient. Temperature advection simply means the rate at which air will cool locally as the result of the wind blowing colder air toward warmer air (cold advection) or warmer air toward colder air (warm advection). As temperature advection takes place, atmospheric pressure in that column of air is altered and therefore the balance between pressure gradient and Coriolis forces that accompany these pressure changes. Imbalance between these two forces requires the wind speed to change such that the Coriolis force comes into balance with the pressure gradient force. This process is always occurring wherever there is temperature advection. *This is the basic mechanism for cyclone formation poleward of about 30° latitude.* Conversely, temperature gradients equatorward of 30° (about half of the earth's surface) are generally small and therefore cyclone formation is relatively rare at those lower latitudes with the exception of hurricanes, which are fueled by cumulus convection. For our purposes it is therefore important that we now look closely at temperature advection and the advection process in some detail.

Advection is likely to be strong when both the wind speed and the temperature gradient along the direction of the wind are large and will be zero when either quantity is zero or when the winds blow at right angles to the gradient; that is, parallel to the isotherms, neither toward colder air nor toward warmer air. The stronger the gradient and the stronger the wind speed directed up or down that gradient the stronger the advection.

As an example, consider the gradient concept illustrated in the previous chapter in conjunction with Figure 1.2. Instead of considering the pressure gradient, let's rework the idea for temperature. Imagine that the temperature in Pittsburgh is 20°C and Philadelphia is 10°C, a difference of 10°C over 500 km (500,000 m). In this case, advection will be positive (warm advection) if the wind direction is from Pittsburgh toward Philadelphia, and negative (cold advection) if the wind is blowing in the opposite direction, from Philadelphia to Pittsburgh. Let's say that the wind speed blowing from Pittsburgh to Philadelphia is 10 m/s.

The magnitude of the advection (warm in this example) would be the product of the wind speed along the direction from Pittsburgh to Philadelphia (10 m/s) times the magnitude of the gradient, which we showed previously in the case of pressure to have the value of 5 mb, the change in pressure between the two points, divided by 500,000 m; temperature advection would be 10°C divided by that same distance. Although advection will be the only quantitative concept introduced in this book, for those who like calculations the magnitude of the advection in this example would be 10 m/s times 10°C (the temperature difference between the two stations) divided by 500,000 m, or 0.0002°C per second. All other things being equal, this means that if temperature advection were the only factor in affecting a local temperature change, the latter would be equal to about 20°C per day (a day containing about 100,000 seconds)—a fairly large value. The larger the wind speed up or down the gradient, the larger the magnitude of the advection. We will, however, not bother to calculate or even quantitatively evaluate advection in subsequent discussion, although a qualitative understanding of the concept and a qualitative assessment of its relative magnitude on a weather map is essential for understanding subsequent material.

Visual estimation of temperature advection

Temperature advection can be easily assessed by a casual inspection of the temperature and pressure fields, such as shown in Figure 2.2. Recall the magnitude of the temperature advection is simply proportional to the product of the magnitude of the temperature gradient times the speed of the wind directed down or up the gradient, which is to say that the wind direction is at right angles to the isotherms blowing either toward higher or lower temperatures, respectively, representing cold- or warm-air advection. Now, the magnitude of the temperature gradient is simply inversely proportional to the spacing of the isotherms. The magnitude of the wind speed is inversely proportional to the spacing of the isobars, which lie parallel to the (geostrophic) wind directions. Advection therefore will be a maximum where the isobars and isotherms are at right angles to each other. This configuration where the two fields intersect at right angles forms a more or less rectangular area, which we refer to as an *advection box*. Advection boxes can be seen on Figure 2.2, examples of which are shaded—one near the letter W, representing warm advection, and one near the letter C, representing cold advection. *The smaller the advection box, the larger is the magnitude of the advection.* Thus, the smallest (warm) advection boxes (the largest advections) are found near the letter W, where the air is blowing from warmer to colder temperatures, and the strongest cold-air advection is near the letter C (signifying cold advection), where the air is blowing from colder to warmer temperatures.

To assess the magnitude of the temperature advection one need only locate advection boxes on a map showing isobars and isotherms. The strongest advections take place where the boxes are smallest. Small advection boxes are meaningful only when compared with large advection boxes (or to an area where no advection boxes exist), since size and number of advection boxes depends on the choice of the isobar and isotherm intervals (e.g., 4 mb and 5°C, as in this case). In fact, the absence of advection boxes does not necessarily signify zero advection but only that the temperature advection is small compared to the smallest advection boxes. Whether the boxes represent cold or warm advection is something determined by inspection. The reader is invited to locate advection boxes on Figures 2.8, 2.9a, and 2.13.

Advection and cyclone development

Let us return to the present example, shown in Figures 2.1 and 2.2. East of the cold front and south of the warm front is an area known as the warm sector, delineated by the upside-down L-shaped configuration of the two intersecting fronts. The warm sector is characterized by warm air moving up from lower latitudes, relatively small temperature gradients, and weak winds. Hence, the warm advection is weak in this sector, as compared with the much stronger warm advection just north of the warm front.

In Chapter 1, we point out that whenever the temperature changes, as would occur with warm and cold advections, the pressure is also obliged to change in response to the change in the density and weight of the column of air. Accordingly, when the temperature changes due to advection, the pressure —and therefore the pressure gradient—must somehow change to accommodate the new weight of the air column. In turn, this would require that the Coriolis force react to this temporary imbalance of forces in order to achieve balance, as was illustrated in the bonfire example of Chapter 1. That figure emphasizes that convergence and the spin up of counterclockwise (cyclonic) rotation occurs in the vicinity of surface pressure falls (a negative pressure tendency) due to the readjustment of the Coriolis force; in the case of pressure falls, the adjustment causes convergence at the surface, upward vertical motion, and (potentially) precipitation in the convergent region. Similarly, divergence and pressure rises tend to take place in regions of cold advection, giving rise to clockwise (anticyclonic) motions and downward vertical motion. These areas tend to be nearly devoid of clouds, except possibly for fair-weather cumulus (see Chapter 3 for a discussion of this cloud form).

We now relate these principles to the weather map sequence shown in Figures 2.1–2.3, recognizing that the correspondence between warm advection, surface pressure falls, and upward vertical motion (and between cold advection, surface pressure rises, and downward vertical motion) is not strictly a one-to-one relationship, as there are always some exceptions to this rule. These relationships do hold in a broad sense as can be seen in these figures.

Before proceeding with our discussion, an important point to reiterate is that, when speaking of the vertical motion in the context of these large-scale processes, we are referring only to the vertical component of the nearly horizontal sideways ascent or descent. One complication to the rule that states that convergence, surface pressure falls, and upward vertical motion are closely related is that warm-air advection, being associated with upward vertical motion, will be partially offset by the expansion cooling associated with that promoted by vertical ascent. Similarly, cold-air advection will be *partially* offset by the compressional warming due to descent. However, observations show that warm-air advection is usually only partially offset by cooling due to ascent. Vice versa, for cold-air advection that, although accompanied by downward vertical motion, is only partially offset by the compressional warming due to descent. As a rule of thumb, the actual local cooling rate is typically about two-thirds the magnitude of the temperature advection at that location.

Exceptions to these relationships among advection, pressure tendency, and vertical motion do occur, however, and these will be briefly discussed later in this chapter via the anomalous case that occurs during the occlusion stages of cyclogenesis. For now, however, let us adopt these relationships among advection, pressure changes, vertical motion, and local warming or cooling as fairly reliable rules of thumb.

Not surprisingly, therefore, extensive layer-type clouds and precipitation resulting from the sideways slant ascent would be expected to occur on the cold side of the warm front where the wind flow is most strongly crossing a field of closely packed isotherms from warm to cold, which is the definition of strong warm-air advection. Indeed, we see this plainly in Figure 2.3.

To return to the surfboard analogy, the surface board is slanted upward more steeply along the direction of air motion just north of the warm front than in the warm sector; it is sloped downward more steeply just west of the cold front where strong cold-air advection and descent is occurring (Figure 2.2). (Here, the surfboard analogy of Chapter 1 must be modified in the analogous case, that of the descending airstream west of the cold front because the soccer ball would have to roll downward when submerged below the surf board. That would only happen if the ball were denser than water,

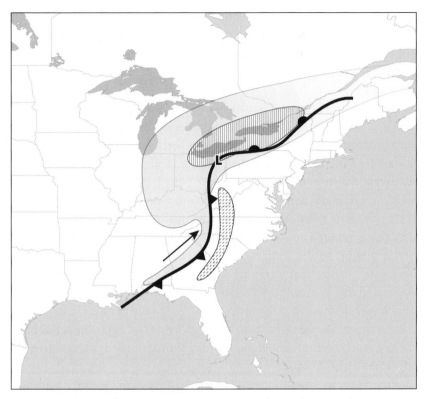

FIGURE 2.3. Clouds and precipitation patterns corresponding to the same schematic weather pattern shown in Figures 2.1 and 2.2 but highlighting the comma-shaped cloud pattern. Light shading indicates overcast cloud cover and hatching denotes steady precipitation. The salient of cloud-free air west of the cold front, dry tongue, is indicated by an arrow.

say if it were filled with lead to simulate the sinking of denser cold air, with the surfboard below the lead ball.) Because of the ascent, the area just north of the warm front is characterized by layers of clouds, and sustained precipitation (Figure 2.3).

For the most part, the dense cloud cover—the shaded area—corresponds to the area of warm advection, where air is crossing isotherms from warmer to colder air, the area located just north of the warm front marked by the letter W in Figure 2.2. The warm front therefore serves as boundary to the state of the weather, acting as the so-called liftoff point where the southerly airstream begins to ascend as it moves northward from the warm sector and overrides the colder air below.

The warm sector

The terminology used here—fronts, warm sector, etc.—originated from Scandinavian meteorologists during the First World War who adopted the wartime terminology that referred to the movement of battle lines. Although some warm advection exists in the warm sector, where the temperature gradients are generally weak, this area generally consists of a variety of clouds and little precipitation resulting from only a slight slantwise ascent.

In the warm sector, air trajectories are almost parallel to the cold front, at least near the cold front, so that the warmest and moistest air (in this example, from the Gulf of Mexico) tends to be found just ahead of the cold front. Although not part of the present discussion, it is worthwhile in light of later discussion pertaining to atmospheric stability to point out that lines of thunderstorms, often quite severe, may be found in the warm sector, lined up and almost parallel to the cold front (and sometimes extending north of the warm front), as shown by the narrow band of precipitation just ahead of the cold front in Figure 2.3.

Vigorous but less violent showers often occur just behind the cold front in a narrow band (the cloud band west of the front in Figure 2.3) where the warm, moist air in the warm sector is being lifted. Many elementary meteorology books often attribute the narrowness of the precipitation and the often vigorous showers that accompany the cold-front passage to a steep slope of the front. Although cold fronts do have a steeper slope than warm fronts over the bottom few hundred meters, the reason for this narrow strip of showers is that the warm air in the warm sector is generally moving almost parallel to the cold front and not rising over the cold air, except in a narrow band along it. Ascent along the warm front, however, occurs over a broad region. Meteorologists often refer to this process of warm air ascending over the cold as overrunning.

The overall cloud and precipitation pattern shown in Figure 2.3 has the appearance of a large comma, with the comma tail more or less corresponding to the cold front and the comma head the region of cloud produced mostly by overrunning north and northwest of the warm front (and by frictionally induced convergence, most importantly over the western part of the comma head; we will treat frictional convergence later in this chapter). Meteorologists often refer to the pattern shown in this figure as the comma-cloud pattern, characterized by a large comma head and a narrow tail, with the former expanding and the latter shrinking during storm development. We will make frequent references to this classic, characteristic shape of the cloud pattern, typically associated with developing and mature cyclones.

The way surface highs and lows move

Lows and highs generally move eastward with the prevailing westerly flow, although often with a poleward or equatorward component, as we will show later in this chapter. Since these highs and lows give the appearance on successive weather maps of continuous motion, it is natural to think of them as discrete entities, like sticks floating in a stream of air. Downstream, the stick is the same stick we saw upstream and the same one that moved through our field of view. Highs and lows, however, do not move in this fashion but continuously reconstitute and destroy themselves as they go from one location to the next.

To clarify, let's look at the pressure tendency pattern (Figure 2.4) associated with our representative weather system, shown in Figures 2.1–2.3. Here, the dashed lines represent the loci of constant pressure tendency, technically referred to as isallobars, labeled in units of millibars change over the previous 3 hours. In order to visualize the wind direction, pressure pattern, and advections, refer to Figures 2.1 and 2.2 and recall the higher pressure to the right of the wind direction, with speed inversely proportional to the spacing of the isobars.

The first thing to note in Figure 2.4 is that the significant pressure rises and falls are relatively confined to two locations: one of the rises is west of the cold front and one of the falls is north of the warm front. Both areas of significant pressure changes correspond to areas of strong advection and small advection boxes, cold and warm, as can be seen from Figures 2.1 and 2.2; both centers of maximum advection are situated just on the cold side of frontal boundaries and not far from the low pressure center. Maximum pressure rises correspond closely with strongest cold-air advection; maximum pressure falls occur in the region of strongest warm-air advection (see Figure 2.2).

A couple of other features of the pressure and pressure tendency patterns in the vicinity of fronts are important to note. First, the tendency lines are discontinuous at the fronts, similar to the discontinuity in temperature gradient at the front. Moving from east to west across the cold front, the pressure falls just ahead of the cold front, then rises on the west side of the front. Similarly, moving from the warm sector northward across the warm front, the pressure tendency changes from weakly negative to strongly negative in accordance with the patterns of temperature advection. In both cases, pressure tendency changes rapidly with a frontal passage.

This abrupt change in pressure tendency at the front is consonant with a pressure minimum at the front, as is evidenced by the noticeable V-shaped kinks in the isobars at these boundaries (see Figure 2.1), making it relatively

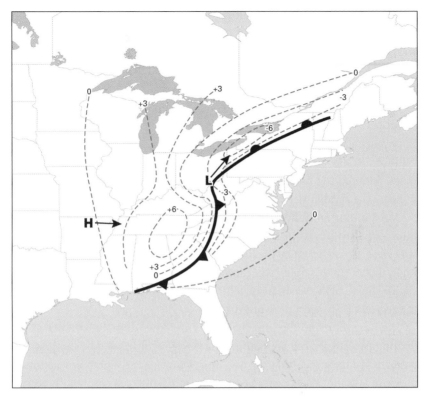

FIGURE 2.4. Same schematic weather system depicted in Figures 2.1–2.3, showing, besides the same frontal boundaries, the contours of pressure tendency over the previous 3 hours (dashed lines) in units of mb per 3 hours. Short, heavy arrows denote the direction of motion of the low and high pressure centers, respectively, toward the region of maximum pressure falls and toward the region of maximum pressure rises.

easy for the analyst to identify the exact location of the front in the isobar pattern.

Further, the passage of a cold or warm front is usually marked by abrupt changes in cloud cover, wind direction, and weather. Like pressure tendency, wind direction also is discontinuous at the fronts, changing from southerly to a northerly and westerly component moving from east to west of the cold front, and from southerly with a westerly component to southerly with an easterly component moving from south to north of the warm front. These changes, however, sometimes occur gradually, in stages, as discussed in Chapter 4.

That the weather will change at the fronts can be seen in Figure 2.3. Brief showers often accompany the passage of a cold front, followed by clearing, whereas a transition to layer cloud and overcast conditions with precipitation may accompany the passage from the warm sector to a location north of the warm front or vice versa. However, the changes are much less dramatic in crossing the warm front than the cold front, although a large area of continuous precipitation tends to be found just poleward of the warm front.

A sudden change in the barometric pressure and/or a change in wind direction and sky cover, therefore serve as indicators of a frontal passage. Fifty years ago, when meteorological station data were sent by teletype, the symbol FROPA (front passed) was often inserted into the code. Weather maps denoted the passage of a front using the check mark symbol ✓ to indicate the sudden rise in pressure following previous pressure falls.

Most importantly, however, the low must move toward regions of maximum pressure falls while the highs move toward regions of maximum pressure rises, as suggested by the direction of the arrows in Figure 2.4. Accordingly, the low in this figure will track toward the northeast, moving on a path slightly north of the warm front, while the high tracks southeastward toward the region of strongest pressure rises.

We now return to the idea that highs and lows do not move like sticks floating in a stream but are continually being reconstituted by the pressure pattern. We have seen that the low moves toward the region of maximum pressure falls, but that occurs because the atmosphere is continually reconstituting a new low—in effect, digging a hole in the pressure pattern ahead of the low and filling it up behind. Likewise, the high will move toward a region of maximum pressure rises because it is being constantly reconstituted by the pressure field.

A somewhat facetious analogy would be as if a hole were incorrectly excavated on the south end of a field and two men were asked to move it to its correct location at the north end of the same field. To accomplish this task, one man proceeds to dig out the northern end of the hole while the other fills in the southern end. Eventually, the hole migrates to the northern end of the field. Although this would be an extremely tedious way to move the hole, it is the way the atmosphere does it. Moreover, should the person on the northern end of the hole dig faster and deeper than the person filling up the southern end, the hole would proceed to widen and deepen as it moves northward.

A simple explanation as to whether the low pressure center will intensify (the central pressure will decrease) or simply move without intensification will depend on the magnitude of the pressure falls near the low center; if these are sufficiently large, the low will deepen. It may even reform elsewhere, appearing to jump from one place to another, depending on the pattern of pressure rises and falls. And these depend, in part, on the temperature advections, which in turn depend on the horizontal temperature gradient and the wind. (Topography can also play a role in influencing pressure changes, irrespective of the temperature advections.)

It is not surprising that low pressure centers often deepen explosively where a strong temperature gradient exists, favoring strong temperature advections and rapid changes in surface pressure; as we will shortly discuss, however, highs never change very rapidly. A reason for this unsymmetrical behavior of highs and lows will be discussed in more detail later in this chapter.

The upper atmosphere

A casual observer of an internet weather site that shows airflow at upper levels in the atmosphere would note that unseasonably warm temperatures occur when the jet stream is to the north and unseasonably cold when it lies to the south (in the Northern Hemisphere). In this section we will begin to discuss the relationship between the flow at high levels, the belt of fast-moving air popularly called the jet stream, and the surface pressure and temperature patterns. In discussing the upper wind patterns we will refer to contours of geopotential height (the height of the isobaric surfaces) as isobars. Although an approximation, these lines designating the heights of the isobaric surfaces are closely analogous to isobars on constant height surfaces. Sea level is one such constant height surface but other levels in the atmosphere can be referred to. For example, lines of geopotential height at the 500-mb pressure surface correspond closely to the isobars on the 5500-m surface. For readers not familiar with geopotential height concept, reference to isobars on a constant height surface instead of geopotential lines on an isobaric surface may be more easily understood.

Above the first couple of kilometers the temperature and pressure patterns differ markedly from those at the surface. Highs and lows (anticyclonic and cyclonic vortices) are still to be found but with some profound differences. Midlatitudes in both hemispheres are characterized by a prevailing westerly wind whose strength increases with height as far as the tropopause, nominally about 12–15 km at midlatitudes (Figure 1.9). In reality, the jet stream is hardly a single core of fast-moving winds but a wide zone of strong westerly winds that contains multiple centers of higher wind speeds, situated within narrow bands, perhaps a few hundred kilometers wide with maximum wind speeds typically 50–75 m/s (100–150 kt). Sometimes these wind speeds are much higher; the local and elongated wind maxima are referred to as jet streaks, and the maximum wind centers are referred to as jet cores. Not uncommonly, the upper jet streak is associated with some sort of surface perturbation. As we will continue to point out, these jet steaks are physically and closely related to horizontal temperature gradients in the atmosphere below them, tending to occur above regions of strong horizontal temperature gradients. Since temperature gradients also tend to weaken with height up to the middle troposphere, jet streaks will be closely associated with strong horizontal temperature gradients near the surface.

It is often observed that the belt of strongest winds, the jet stream, splits along the West Coast of the United States, with one branch bending southward over the states bordering the Gulf of Mexico and one branch remaining over the northern part of the United States before they join along the East Coast of North America. As just noted, and to be discussed further, the jet stream tends to be located close to the region of maximum surface temperature gradients, with lower temperatures to the left of the jet, and higher temperatures to the right of the jet along its direction of motion. So it is not surprising that a branch of the jet stream might favor a westerly direction along the northern border of the Gulf of Mexico or in a southerly direction along the East Coast in winter, where strong horizontal temperature gradients exist between the warm waters and the land.

As the jet stream is in close geostrophic balance, higher pressure being on the right side of the flow and lower pressure on the left, the pressures on constant height surfaces will generally be higher to the south and lower to the north of the isobars in Figure 2.5, except in the vicinity of actual centers of rotation. The small ellipses in this figure denote the presence of the jet cores, regions of locally stronger winds.

Highs and lows also exist at upper levels but in a different form than the cellular structures we see on the surface weather maps, such as in Figure 2.1. Superimposed upon the prevailing westerly direction are cyclonic and anticyclonic vortices, centers of locally lower and higher pressure, respectively. Unlike their counterparts at the surface, these centers of rotation do not resemble the cellular patterns found there in that they usually do not correspond to closed isobars, although the two features are related. When a vortex, cyclonic or anticyclonic, is superimposed on a relatively uniform westerly current of air, such as exists at levels of 3–15 km above the surface, the westerly current and vortex manifest themselves as wavelike undulations, with troughs and ridges corresponding, respectively, to locally lower pressure (the embedded cyclonic vortex) and locally higher pressure (the embedded anticyclonic vortex).

In other words, the strong westerly current masks the upper cyclonic and anticyclonic vortices, whose centers are marked by a pair of Xs, respectively, along the trough and the ridge axes in Figure 2.5. Note that the centers of rotation do not necessarily coincide with the centers of low and high pressure; in this case, the actual centers of rotation lie in the trough or ridge axes, not in the low or high pressure centers located farther north within the trough (embedded cyclonic vortex) or farther south within the ridge (embedded anticyclonic vortex). Instead of a distinct cellular pattern, close inspection will show that these centers of rotation manifest themselves at these levels as centers of enhanced cyclonic or anticyclonic curvature of the isobars. In effect, they constitute hidden centers of rotation. Like the jet streaks, there can be multiple smaller vortices embedded within a larger-scale trough or ridge. Meteorologists measure the degree of rotation using a quantity known as vorticity, whose relative maxima and minima are shown in this schematic figure by the points marked X, denoting a maximum of cyclonic rotation in the trough and a maximum of anticyclonic rotation in the ridge.

Figure 2.5 shows the trough and ridge system associated with the high and low pressure pattern shown in Figures 2.1–2.4. Note that the trough, the upper cyclonic vortex embedded in the westerly stream, lags to the west of the surface low pressure center and the ridge lags to the west of the surface high pressure system, which is not shown here but is analogous to the ridge at upper levels to the east of the surface low—off the edge of our maps. Since the jet stream coincides with strong lower-troposphere horizontal temperature contrast, it also corresponds to the belt of storms and fronts. Areas poleward of the jet stream (to the left of the wind flow) are colder in the lower troposphere than areas equatorward of the jet stream (to the right of the wind flow). Areas east of the trough axis and near the location of the jet stream generally correspond to areas of clouds and precipitation embedded in the southerly flow, while areas to the west of the trough (and east of the ridge) correspond to areas of fair weather and rapidly descending northerly flow, as indicated in Figure 2.6. Within the center of the trough (poleward of the jet stream), the cold air is often accompanied by showers due to decreased atmospheric stability; the latter concept of static stability will be discussed later in this chapter. Although we have chosen not to depict them, local jet cores, as shown in this figure, also correspond to local centers of cyclonic and anticyclonic rotation on their flanks.

This westward displacement of the ridge and trough lines and the centers of anticyclonic and cyclonic rotation aloft from the surface high and low pressure systems is, contrary to some elementary meteorology texts, not

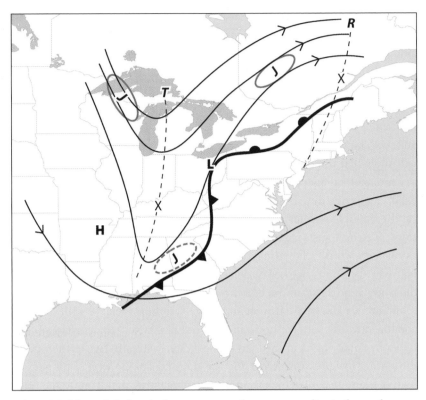

FIGURE 2.5. Schematic isobars in the upper troposphere corresponding to the weather patterns shown in Figures 2.1–2.4. Contour lines represent isobars but are not labeled, as it is understood that higher pressure lies to the right (equatorward) of the contours, along which the winds are approximately geostrophic. The wind barbs along the contours are intended to denote the direction of the winds blowing parallel to the isobars. The pair of Xs represents centers of the major cyclonic and anticyclonic vortices along the respective trough and ridge axes, respectively labeled T and R. Surface fronts and the surface high and low pressure centers are also marked. The elliptical shapes labeled J denote the location of two jet cores, plus a third one (dashed) indicating the formation of a new jet core.

a tilt of the surface features but an actual westward displacement from the surface low or high at levels above the first 2 or 3 km. This lag is essential for a temperature gradient to occur and for temperature advections to take place near the surface fronts. It is therefore also an inherent necessity for cyclone movement and development. Indeed, cyclone development and surface temperature advection are maximized when the lag between the surface low or high and the upper trough and ridge is about one-fourth to one-half the distance between trough and ridge, and they are weak when the lag is zero. We will return to this very important feature of the weather pattern and show how the westward lag of the upper trough from the surface low is directly related to cyclone intensification and its absence in the cessation of that cyclone development.

Where do the cold and warm airstreams originate and where do they go?

These features represented in Figures 2.1–2.5 are manifestations of the exchange of warm and cold air across latitudes resulting from the latitudinal heating imbalance. Were this exchange not to take place, the tropics would heat up and the higher latitudes would cool down indefinitely. While it is evident in Figure 2.1 that cold air is moving southward and warm air is moving northward on either side of the trough axis, the two-dimensional representation of this exchange is limited, as the air is actually moving in a third (vertical) dimension, albeit at a very small angle with the horizontal.

We can see this exchange more clearly when represented in two different ways: in three dimensions in Figure 2.6a and as a cross section in Figure 2.6b. In the former, the most rapid rise (represented by the tubular-looking trajectory of the airstream) occurs in the region of greatest pressure falls and continuous precipitation. Of course, since most of the moisture is found in the lowest couple of kilometers, the ascent is most critical for precipitation when it is occurring in the lower troposphere, which is a distance not far north of the warm front. Similarly, the driest air descends west of the cold front, as represented by the appropriate tubular trajectory in that area, originating west of the upper trough (Figure 2.5). It must be emphasized, if it isn't already obvious, that the trajectories depicted in Figure 2.6 represent only two of a family of streamlines below, above, and on either side of the ones represented.

Although the actual airflow is more complicated than depicted in this figure (as a single pair of tubes representing the material trajectories of the air), we can see that the air in the warm sector rises slowly until it reaches the warm front and subsequently ascends rapidly into the upper troposphere, where it joins the westerly current. Similarly, a branch of northerly flow, originating in the westerly current west of the upper trough and in the high troposphere, peels off toward the south and descends into the lower troposphere just east of the high pressure center and west of the surface cold front.

Figure 2.6b shows the flow from south to north along a two-dimensional slice coinciding with the ascending trajectory in Figure 2.6a. Streamlines are shown rising most rapidly just poleward of the warm front, with the strongest ascent near middle levels (6–8-km elevation). It is implied in Figure

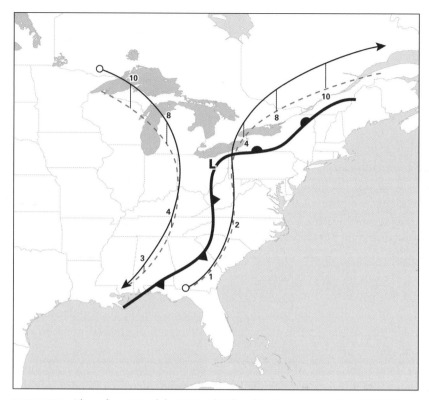

FIGURE 2.6A. Three-dimensional depiction of airflow from two source regions: the high troposphere west of the trough axis and the southern end of the warm sector for the same schematic weather pattern shown in Figures 2.1–2.5. Numbers aside the trajectory indicate the altitude of the airstream in kilometers. The dashed lines represent the projection of the trajectory on the surface.

2.6a, that the air reaching upper levels above the frontal zone in Figure 2.6b is turning into the page toward the reader, whereas ascent in and below the frontal zone occurs in air that originates north of the warm front and moves with an easterly (away from the reader) component as it ascends.

Many elementary books on meteorology tend to show the retreating cold air north of the warm front, often colored blue to depict its lower temperatures, as a shallow wedge over which the warm air flows from the south. In fact, the transition from colder air near the surface to overrunning warm air aloft occurs over a frontal zone whose boundaries are represented by the sloping thin pecked lines in Figure 2.6b. Cloud is formed once the rising air from the warm sector and the more gently rising air in the retreating cold air reach saturation. As previously noted, most of the precipitation originates

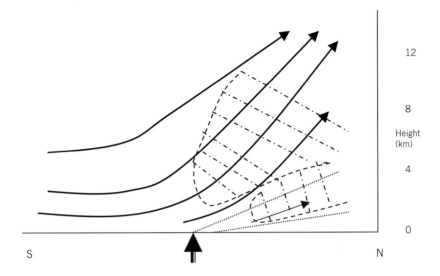

FIGURE 2.6B. Schematic cross section (height versus distance from south [S] to north [N]) along the tubular trajectory originating east of the cold front in Figure 2.6a. Barbed streamlines (and the thin arrow) depict the wind flow in that plane. Thin pecked lines show the sloping warm frontal zone, dashed cross hatching denotes cloud, and the heavy solid arrow below indicates the surface position of the warm front.

in cloud (hatched areas) formed where the air is rising most rapidly in the lower troposphere. As suggested by Figure 2.1, the airflow within the retreating cold air below the frontal zone has a generally southeasterly component.

Figure 2.6 retains only the basic airflow; it does not tell the entire story. It is meant to show that troughs and ridges viewed on conventional weather maps are two-dimensional manifestations of three-dimensional latitudinal exchanges of air. Further elaboration on the nature of these airstreams follows in the next section.

Vertical temperature profile: inversions

One might wonder upon looking at Figure 2.6a how the so-called cold air descending west of the cold front can continue to be called *cold* and the so-called warm air ascending from the warm sector continue to be referred to as *warm* after experiencing considerable vertical displacement, whereby rising warm air cools by expansion and descending air warms by compression. After all, the descending air experiences a warming due to compression of about 10°C for each kilometer of descent. Similarly, ascent of warm air will rapidly cool by the same amount, unless condensation takes place, in which case the air still cools but, because of the release of latent heat of condensation, at a reduced rate—typically about 5°–6°C per kilometer in the lower troposphere.

Let us look at this seeming contradiction a little closer. Because the ground is a hard surface, air does not move through it. This constraint requires that the vertical motion be virtually zero at the surface. Away from the constraining lower boundary, the air is free to move upward or downward, so that it actually reaches a maximum value in its vertical motion somewhere in the middle troposphere, typically at an altitude of about 6 or 8 km, above which the magnitude of the vertical velocity decreases with height. Consequently, the temperature change of ascending or descending air near the surface will be much smaller in magnitude than in the middle troposphere. Thus, the northerly flow of cold air originating near the surface east of the high pressure system over Canada undergoes only a small component of descent, so that it retains its cold attribute. In contrast, the descending airstream shown in Figure 2.6a, whose origins are in the upper troposphere, experiences considerable vertical descent and therefore undergoes considerable warming while bringing with it extremely dry air. We can imagine with a glance at Figure 2.6a that a cold airstream (not depicted in this figure), originating near the surface over Canada, follows a trajectory similar to the projection of the tubular arrow on the surface (the dashed line) but arrives west of the cold front *beneath* the more rapidly descending airstream while remaining cold.

Air descending from high levels, however, undergoes considerable compressional warming. We illustrate the resulting temperature profile for the descending, midtropospheric air, as well as for the two ascending airstreams east of the cold front, with the aid of the three vertical temperature profiles (temperature versus height) shown in Figure 2.7a. One sounding is for a location just west of the cold front within the cold air but also within the rapidly descending airstream shown in Figure 2.6a. A second sounding illustrates the temperature profile in the warm sector. A third is located just north of the warm front.

This first sounding is characterized by a temperature inversion in the lower troposphere, in which the temperature actually increases with height within a shallow layer formed by this sinking and rapidly warming airstream. The descending airstream in Figure 2.6a, having warmed due to compression as it sinks, is located above the inversion, capping the cold air whose surface origin is at low levels over Canada, having experienced only a small descent.

In contrast, the ascending airstream rises very rapidly north of the warm front, where it reaches the upper troposphere, joining the upper westerly

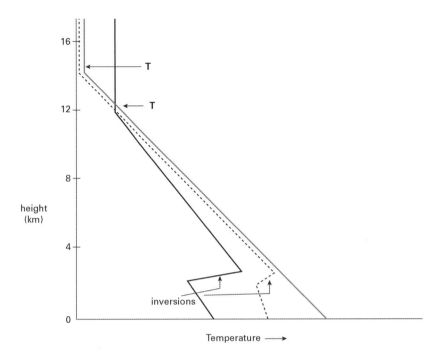

FIGURE 2.7A. Schematic representation of three vertical temperature profiles (temperature versus height): within the warm sector (thin solid line), just north of the warm front (dashed line), and west of the cold front (heavy solid line). Temperature inversions are indicated for two of the soundings. The tropopause is also marked with the letter T, showing that this feature is at a much lower level in the descending part of the flow.

flow near the downstream ridge; this turning of the flow toward the east is shown in Figure 2.6a but can not be shown in Figure 2.6b. Continuous precipitation, either as rain or snow, marks the trajectory's rapid rise from low levels north of the warm front over the cold-air wedge, which is identified below the inversion in the second sounding located north of the warm front in Figure 2.7a. Air originating within the warm sector arrives above this cold air, forming an inversion between the cold air near the surface and the overrunning warm air; the interface between these two airstreams forms a sloping frontal zone. The latter is marked by the pair of thin sloping pecked lines in Figure 2.6b, which meet the surface at the location of the surface warm front. The inversion formed by the overrunning warm air, is located within these sloping pecked lines, and is shown in the lower part of the thin dashed sounding in Figure 2.7a. Later in this chapter, we will return to the importance of the temperature soundings in the development of cyclones.

Unlike the sounding west of the cold front, the inversion identified in the sounding north of the warm front is caused by a very different process than in the profile west of the cold front; that inversion is formed by the overrunning of warm air from the warm sector above the colder air below rather than by strong sinking motion. Air within the warm sector flowing northward arrives at the warm front and then overruns the cold, retreating airstream, as depicted in Figure 2.6b. In this figure the transition between the shallow cold-air wedge and the ascending warm air is marked by pecked lines (which is actually a shallow transition zone) constituting the top of the warm front, which slopes backward toward the cold air with height, thereby forming the temperature inversion and the lower boundary of the ascending warm air. Below this inversion, cold air near the surface is also ascending, while moving westward and northward from a location farther east along the front; its probable origin in this schematic example would be over the ocean or the Canadian Maritime provinces in this particular situation.

The third sounding, that of the vertical temperature profile located within the warm sector, shown in Figure 2.7a, shows the greatest decrease in temperature with height of the three soundings, largely because of the presence of warm air at low levels.

All three soundings in this figure exhibit only small temperature differences at middle levels, although the air west of the cold front remains somewhat colder than that in the warm sector. Small temperature differences in the middle troposphere result from the vertical motions in which descending (and warming) airstreams west of the surface cold front, and rising (and cooling) airstreams east of the surface front and north of the warm front eliminate the strong horizontal temperature contrast so evident near the surface. Strongest temperature gradients and temperature advections are therefore found near the surface and these, like the temperature gradients, weaken rapidly with height up to the middle troposphere. For this reason, surface temperature fronts do not usually extend above the lowest few kilometers, as is shown by the sloping warm front in Figure 2.6b and by the shallowness of the warm-front inversion, also shown in Figure 2.7a.

Tropopause: the weather's cap
The tropopause puts a cap on our weather systems. Vertical exchanges of air, depicted in Figures 2.6a and 2.6b, are mostly confined below the tropopause, which resembles a temperature inversion serving to put a brake on vertical motion. As shown in Figure 2.7a, the temperature ceases to decrease with height, becoming nearly constant with height just above the tropo-

pause, marked by the letter T. The air just above the tropopause is called the stratosphere, which extends several kilometers in the vertical. Because stratospheric air does not readily mix with the troposphere, almost no water vapor is contained in the stratosphere. High concentrations of ozone gas are also found in the stratosphere, the result of dissociation of oxygen molecules by the impaction of photons from the sun, which occurs well above the tropopause. Volcanic eruptions often push their debris into the stratosphere, where the material is carried great distances. During the 1960s and 1970s, atomic bomb tests tended to lift radioactive debris into the stratosphere, only to fall out at locations far from their source. One such fallout episode, resulting from atom bomb tests in China, occurred during the 1970s in State College, Pennsylvania.

The tropopause height varies considerably with the tropospheric weather patterns. It is typically lower in the troughs and higher in the ridges. It is lifted in ascending airstreams and lowered in descending airstreams. Unlike the temperature gradients at middle levels and similar to those at the surface, temperature gradients near the tropopause tend to be large. Above the level of maximum winds and very close to the tropopause the direction of the gradient becomes reversed with cold air to the south and warm air to the north. This latitudinal reversal of the temperature gradient can be seen by comparing the three soundings in Figure 2.7a: the air is much colder above the tropopause over the warm sector and north of the warm front than above the tropopause west of the cold front, warm air being found in the trough and cold air in the ridge at these levels. The tropopause therefore tilts downward from high to low, from ridge to trough, and generally from regions of warmer air to regions of colder air at the surface. Since the tropopause also acts as a brake on the vertical motions (the latter being much weaker than at middle levels), strong horizontal temperature contrasts can be maintained near tropopause level without being ironed out by vertical motions.

One can sense, as a passenger on a commercial jet aircraft, the frontlike transition from troposphere to stratosphere. This occurs in level flight when the aircraft enters the stratosphere, as might occur when flying into a trough. In this instance, at least when the passenger is provided with such temperature measurements, the observer will note a sharp rise in temperature and a markedly smoother flight. Conversely, upon entering the troposphere in flying from trough to ridge, temperatures decrease and the air tends to be less smooth.

Intermittent exchanges of air do occur between stratosphere and troposphere, however. In regions of cyclone development, the lower portion of the stratosphere may be dragged down into the troposphere in the region of strong descent, such as in the descending airstream shown in Figure 2.6a. These extrusions of very dry, stratospheric air are found just above the inversion shown in the cold-front sounding in Figure 2.7a; this transfer from stratosphere to troposphere becomes stronger with time as the storm intensifies, eventually dragging very dry air around the trough and then east of the trough axis, creating the cloud-free salient just west of the cold front known to meteorologists as the dry tongue (see Figure 2.3).

Atmospheric stability revisited

In this book, we have discussed two types of atmospheric instability: the sideways or slantwise (baroclinic) instability that affects the large-scale weather patterns (highs and lows) and depends in part on the horizontal temperature gradients; and the small-scale vertical (convective) instability that affects the growth of cumuliform clouds, such as shown in Figure 1.7, and depends on the buoyant force associated with the difference in density between air inside and outside a particular cloud or bubble of air. In turn, the development of convective instability depends on the vertical temperature profile. In this section we will focus more on the latter type of instability.

The ability of the atmosphere to remain stable with respect to locally induced upward or downward motions depends on the rate of change of temperature with height. To restate principles first referred to in Chapter 1, a parcel of air lifted or lowered can be said to be stable if it is pushed back to its original level and unstable if the buoyant force continues to push the parcel upward or downward. If one lifts such a parcel by, say, 1 km, it will cool by expansion, by approximately 10°C. If the atmosphere is colder (more dense) than the parcel of air at the parcel's new level, the parcel will be forced to descend to its original level as the result of a downward force provided by the negative buoyancy (the buoyant force, whether positive [upward] or negative [downward] is closely proportional to the difference between the temperature of the surroundings and that of the air parcel). If the air parcel becomes warmer than the surrounding atmosphere, it will continue to rise until such time as it again reaches a level where the ambient air is warmer than the pocket of air. The latter case is said to be convectively unstable. This is illustrated by the path labeled S in Figure 2.7b.

Figure 2.7b shows a schematic distribution of temperature versus height in which the surrounding decrease in temperature with height, referred to as the lapse rate, is 7°C per kilometer. Paths taken by two parcels of air subject to lifting are shown. The dry air cools at a rate of 10°C per kilometer and

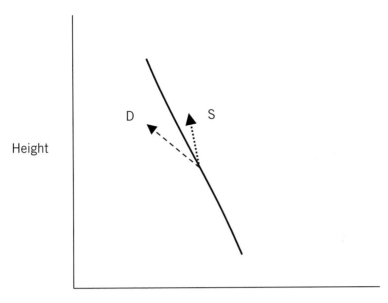

Height

Temperature

FIGURE 2.7B. Schematic distribution of temperature versus height, showing the paths taken by a dry parcel of air (D) lifted and cooled at a rate of 10°C per kilometer, and a saturated parcel of air (S) lifted and cooled at a rate of 6°C per kilometer. The ambient temperature profile, whose temperature decreases by a rate of 7°C per kilometer, is denoted by the heavy sold line. The path S demonstrates convective instability.

therefore remains cooler than the ambient sounding (path D), whose temperature decreases at a rate of 7°C per kilometer. The path labeled S pertains to a saturated parcel of air that cools at a rate of about 6°C per kilometer, thereby remaining warmer than the surrounding air, which cools more rapidly with height. Clearly, the presence of a temperature inversion in which the temperature increases with height, such as shown on two soundings in Figure 2.7a, will cause air pockets to remain stable to either dry or moist ascent. Inversions (or layers in which the temperature is virtually constant with height, such as at the tropopause), therefore, are very efficient suppressers of cumulus convection.

Overall (aside from temperature inversions, such as shown in Figure 2.7a), temperature typically decreases with height at a rate of 5°–8°C per kilometer in the troposphere; the change in temperature with height has been referred to as the lapse rate. In experiencing a 10°C per kilometer rate of cooling, dry air lifted by, say, 1 km, will usually be colder (more dense) than the surroundings (whose temperature would typically decrease by 5°–8°C), arriving with a temperature 10°C below that at its point of origin. It is thereby pushed back downward by a negatively buoyant force resulting from being 2°–5°C colder than the environment. If that parcel of air were to find itself warmer than its surroundings (positively buoyant), it would continue to rise at an accelerating speed, the path labeled S in Figure 2.7b. (Note that cold air will be denser than warm air only when compared at the same level and that the buoyant force is slightly affected by the water vapor content of the air within and outside the parcel.)

Three aspects of Figure 2.7b should be emphasized. First, the steeper the lapse rate (the larger the decrease in temperature with height), the weaker the restraint on the convection. Second, the steeper the lapse rate, the larger the positive buoyant force for saturated air, here illustrated by the path labeled S. Third, when the decrease in temperature with height in the surroundings is less than about 5°C per kilometer convection is prevented from occurring even when the air is saturated, because the temperature of the air parcel will always remain colder than the environment when lifted. The presence or absence of convection is therefore sensitively dependent on the lapse rate of temperature in the surroundings.

Although the atmosphere when assessed on a scale of weather systems is convectively stable with respect to lifting or lowering an air parcel, an exception can be found in the so-called atmospheric boundary layer during the day. The boundary layer is that layer of air adjacent to the ground where vertical mixing of heat and water vapor resulting from surface heating is occurring, as schematically indicated in Figure 1.6; this process typically affects a depth of a few kilometers or less. When air near the ground is heated by the sun, the rate of temperature decrease with height tends to be greater than 10°C per kilometer, at least over a layer close to the ground. The result in this condition is that a pocketful of air lifted from the surface at a rate of 10°C per kilometer will become immediately warmer than the surroundings, and therefore it will continue to rise as the result of an upward (positive) buoyant force. This is illustrated in more detail in Chapter 4.

The atmospheric boundary layer during a sunny day is a roiling mass of invisible rising bubbles of air called thermals, looking somewhat like the bubbles in a saucepan of boiling water, as illustrated in Figure 3.24. Birds and gliders like to take advantage of thermals, which are typically found over terrain heated by the sun. Many of these bubbles survive to achieve condensation as the result of cooling during their expansion, thereby forming the very familiar cumulus clouds, to be discussed in the next chapter.

When condensation occurs in these rising bubbles of warm air, thereby creating a cumulus cloud, the air pockets (bubbles) may continue to be positively buoyant because the rising cloudy bubbles do not cool as rapidly with height as do bubbles of dry air. Condensation thus provides an added boost to the cumulus clouds, though the fair-weather type of cumulus quickly dissipates due to the evaporative cooling of water drops in the cloud, which is caused by the ingestion of dry air from the surroundings. Even midlevel stratiform clouds, such as alto cumulus (discussed in the next chapter), can sprout turrets as the result of a marginal convective instability in the cloud layer.

Inspection of the warm-sector sounding in Figure 2.7a will attest to the fact that the air in this quadrant is less convectively stable than that in the cold air west of the cold front, which, overall, has a smaller decrease in temperature with height and a very strong temperature inversion. In other words, when evaluated over the lower part of the atmosphere, the temperature in the sounding west of the cold front does not decrease as rapidly with height in this sinking airstream as in the warm sector, or even in the sounding north of the warm front. Not surprisingly, therefore, deep convection is more favorable within the warm sector, notably close to the cold front where the air near the surface is warmest and contains the highest moisture content within the cyclonic system by virtue of its origin farthest south, as shown in Figure 3.31.

Large lapse rates, often promoted by strong surface heating, favor the formation of deep convective clouds. As emphasized in our previous discussion, the larger the lapse rate, the weaker the restraining effect of static (vertical) stability on the growth of convective clouds. As discussed later, the atmospheric stability, as measured by its ability to restore local parcels of air to their original levels in the atmosphere when lifted or lowered, is also vitally important not only in governing the appearance and growth of thunderstorms but in restraining or augmenting the growth of large-scale cyclones. As such, it is necessary to understand the concept of atmospheric stability for cyclogenesis, as well as for understanding how thunderstorms form. The former topic will be treated later in this chapter. A further discussion of atmospheric lapse rate and its effect on convection is presented in Chapter 4.

Occluded fronts: a view from space

When the frontal model was first proposed by Scandinavian meteorologists almost 100 years ago, and for some time thereafter, the idea was that the occluded front and occluded low pressure system occurred because the advancing cold air overtakes the retreating cold air along the warm front, thereby lifting the warm air and the retreating cold air to form a kind of triple-decker air sandwich consisting of advancing cold air underneath, retreating cold air in the middle, and warm air on top. In so doing, this process isolated the low pressure center. The front was then called an occluded front and the low an occluded low.

Recent studies have shown that the occlusion process is more complex than initially supposed, much more than simply a matter of the cold front moving faster than the warm front. Rather than a mechanical explanation based on the concept of slow versus fast, the process of occlusion actually involves a more fundamental series of processes occurring over the entire troposphere, whereby the temperature and advection patterns become distorted during the evolution of the cyclone. Instead of delving into the occlusion process, we will simply refer to occlusion as occurring when the cyclone begins to separate itself from the main frontal system and begins to diminish in intensity.

An occluded low and frontal system is shown in Figure 2.8, in which the fronts and isobars are superimposed on a satellite image of a typical comma-shaped cloud pattern. Here, the comma is somewhat more developed than in Figure 2.3, the comma head extending farther west and southwest of the low pressure center in Figure 2.8 than in Figure 2.3. The surface low is now shown cut off from the intersection of the cold and warm fronts, and therefore it is also cut off from the warm sector and from the benefit of strong vertical ascent and low-level convergence, which are inherent in the process of surface pressure falls.

Between the westward bulge of the comma head and the cold front is a narrow, wedge-shaped, largely cloud-free salient, called the dry tongue, shown by an arrow in Figure 2.3. The dry tongue is a sign that some of the dry stratospheric air descending from the high troposphere, depicted by the descending streamline in Figure 2.6a, is beginning to turn in a counterclockwise fashion and, ceasing to descend farther, flows back toward the north in a path that takes it parallel to the cold front and very close to the center of the cyclone. This feature, which constitutes just one of the many changes taking place during the beginning of the occlusion process, tends to be found just southwest or south of the surface low and behind the cold front. Thus the appearance of the dry tongue signals not only the beginning of the occlusion process but also subsequent weakening of the low. Meteorologists tend to regard the development of a dry tongue as the beginning of the end of the storm.

Accompanying the formation of the dry tongue is the process by which the surface low becomes isolated from the cold- and warm-front intersection

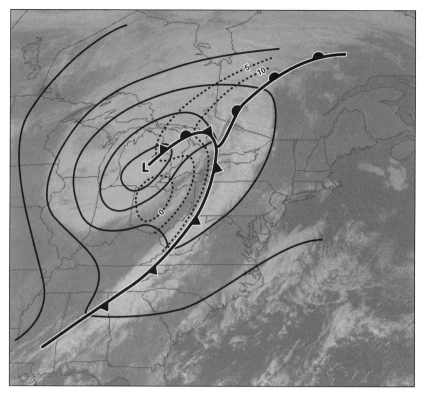

FIGURE 2.8. Visible satellite image of an occluding low pressure system. The comma head and tail are separated by a narrow intrusion of clear air, the dry tongue. Superimposed on the image are the approximate locations of the warm and cold fronts and the occluded front (denoted by barbs and half circles on the same side of the front), the isobars (solid lines), the low pressure center, and some representative isotherms (labeled in degrees Celsius as dashed lines). Note the presence of warm and cold advection boxes located, respectively, northwest of the warm front and southeast of the low center. Courtesy of NOAA/NESDIS.

and moves back into the cold air toward the center of the comma head. Extending from that low center to the intersection of the warm and cold fronts is the occluded front (depicted by points and half circles on the forward side of the front), which is bent back into the comma head.

Occluded fronts have a distinctly different temperature structure than cold or warm fronts. Instead of a discontinuity in the temperature gradient across the occluded front, as is the case for both the cold and warm fronts, the temperature gradients on either side of the occluded front are about equal and relatively weak compared with those behind the cold front. While the temperatures along the occluded front tend to be highest, the temperature advection near the center of the occluded low center becomes weaker as the occlusion process continues. Figure 2.8 shows that a relative maximum temperature is found at the location of the occluded front, with equal gradients on either side. At the same time, a temperature minimum begins to form near the center of the occluded low as shown by the dashed isotherms in Figure 2.8.

The result of this occlusion process is the diminishing temperature advection and weakening vertical motion near and around the low pressure center. As these advections constitute the essential engine for storm development, this change marks the beginning of the storm's demise, at least that of the central low pressure center. In concrete terms, the precipitation diminishes in the vicinity of the occluded low, although wind speeds remain strong around it, as suggested by the closely packed isobars. Further development can occur away from the parent low, such as along the warm front where warm advection is still strong in this figure.

Despite the rather impressive extent of the comma head, much of this cloud is becoming relatively inert with regard to sustained precipitation, which may still continue as showers. Farther to the east, poleward of the warm front, the temperature advection remains strong and the precipitation more sustained and heavy. The business end of the weather system has moved on.

Cyclone intensification (cyclogenesis)

The process by which the central pressure of a low decreases with time is called cyclogenesis. Cyclogenesis is a manifestation of slantwise or horizontal instability, referred to earlier. As the low deepens, the isobars become increasingly closer together, so that not only do the surface wind speeds increase but so do the cold and warm advections that depend on the strength of the winds and the horizontal temperature gradients. Cyclogenesis, like any instability, often occurs rather quickly and then similarly quickly ceases. Cyclones in which the central pressure deepens by more than 12 mb within 12 hours are sometimes called bombs, in colloquial jargon, although this term is hardly very scientific. A somewhat better and more scientific term is explosive cyclogenesis.

An example of such a rapidly deepening low pressure system is shown in Figures 2.9a and 2.9b. In the former, we note that the surface low is situated on the Atlantic coast along a weak frontal wave where precipitation is occurring and a warm sector is just starting to form. Most significant is that the low pressure center is embedded within a strong temperature gradient augmented by the proximity of cold land and a warm ocean surface. North-

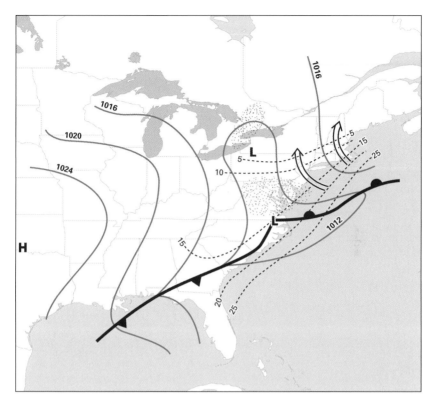

FIGURE 2.9A. Sea-level isobars, fronts, and sustained (continuous) precipitation (shaded areas) for 7:00 am EST, October 14, 2010. Dashed lines labeled in degrees Celsius are the surface-level isotherms, and the double-shafted arrows denote the geostrophic surface wind direction in the region of strongest warm air advection—where the advection boxes are smallest.

east of the low pressure center is a region of strong warm-air advection (the double-shafted arrows representing the direction of the wind across a strong temperature gradient) over New England. Recalling our previous discussion, the strongest pressure falls (dictating the direction of motion of the low) would be expected to coincide with the strongest warm-air advection, represented by the smallest advection boxes.

Hence, it is not surprising in view of the strong temperature advection that 24 hours later (Figure 2.9b) the low pressure center has deepened considerably (by about 20 mb) and has moved in the direction of the strongest warm temperature advection in Figure 2.9a and largest surface pressure falls—toward the southeastern corner of New England. In so doing, the storm experienced a broadening of the precipitation area, and it developed

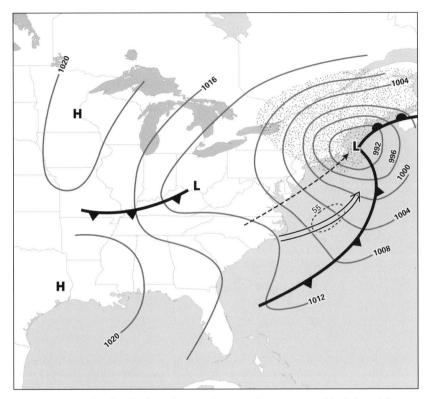

FIGURE 2.9B. Surface-level isobars, front, and sustained precipitation (shaded area) for 7:00 am EST, October 15, 2010. The dashed arrow shows the trajectory of the low center during the previous 24 hours. The curved, double-shafted arrow and the closed contour labeled 55 refer to the location of a newly formed upper (approximately 10 km) jet maximum, whose speed was 55 m/s (about 110 kt).

a well-defined warm sector, resembling the schematic example shown in Figures 2.1–2.4. As most of the pressure deepening at the low center occurred during the first 12 hours after the time illustrated in Figure 2.9a, this storm would have been classified as a bomb; New Englanders might prefer the term nor'easter, notorious for its high winds and heavy snows, the latter occurring most dramatically west of the low center and just inland from the coast. As we will see, two other factors besides strong warm-air advection were important causes of this storm's intensification.

Coastal storms

Coastal cyclones are some of the most damaging storms to affect the East Coast of North America. Snowfall amounts of 2–3 feet inland, especially over hilly areas near the coast, are not unusual in winter; such blizzards are capable of shutting down towns and cities along the coast. The heaviest snowfalls tend to occur north and northwest of the storm center and even west and southwest of the center once the storm reaches its full intensity. This occurs when the strong warm-air advection extends toward the west and north of the low center, causing the comma head to become increasingly distended in these directions.

Because these types of storms tend to hug the coast, the rain–snow line also tends to lie very close to the land–sea border. Snow changing to rain or freezing rain occurs in the areas just on the cold side of the warm front, which also tends to hug the land–sea border. We will return to a more detailed discussion of coastal storms in Chapter 4, and here focus more on the dynamics of storm intensification and motion.

A notable feature of such coastal storms is their almost uncanny ability to follow the coastline, first along the Gulf of Mexico and subsequently along the Atlantic coast as far as the Canadian Maritime provinces. Nearly all cases of explosive cyclogenesis over the northeastern United States are coastal storms; rapid development tends to occur in these storms when the initial disturbance reaches the Atlantic coastline, after which they tend to follow the coast northward in the direction of the strongest warm-air advection.

The greatest potential for development is therefore present when the temperature gradients across the coastline are strongest, a situation that exists during the colder months (October through March), when the sea is significantly warmer than the land. Acting as a channel for the motion of the storm, the strong horizontal temperature gradient and its attendant warm-air advection guide the cyclone along the coast. Even the warm front tends to orientate itself along the coastline north of the storm center, although detailed analysis shows that a weak coastal front is often lurking not far off the coast in wintertime even in quiescent weather conditions.

Figure 2.10 illustrates the sequence of events during the evolution of a typical coastal storm. Initially, the storm forms along a weak frontal zone in the Gulf of Mexico (top left), intensifies during successive 24-hour intervals, and finally begins to occlude (bottom right). The storm reaches its maximum strength just prior to the latest map, accompanied by very strong winds (as evidenced by the close spacing of the isobars) and a large area of precipitation, including abundant snow along the northern and western sides of the low center.

The similarity of the second and fourth sequences (upper and lower right-hand side, respectively) to the maps in Figures 2.9a and 2.9b emphasizes the

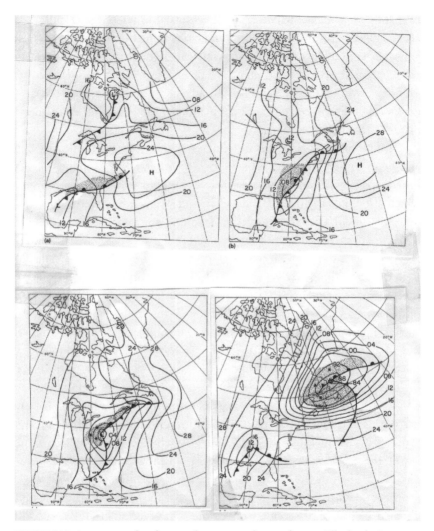

FIGURE 2.10. A sequence of surface weather maps 24-h apart for a rapidly developing coastal storm during December 1981, starting with the upper left, proceeding to the upper right, lower left, and finally lower right panel. Fronts, isobars (labeled in millibars above 1000), and sustained precipitation areas (shading; stars represent snow) are shown. (Carlson, *Mid-Latitude Weather Systems*)

common underlying dynamics governing these types of storms and therefore the general utility of the cyclone model for coastal storms. The next section addresses additional reasons, besides the strong horizontal temperature gradient between land and sea, for such explosive coastal cyclogenesis.

Coastal cyclogenesis, frontogenesis, and the jet stream

We will now try to explain the interrelationship of fronts, cyclones, the jet stream, and coastal cyclogenesis. First, let's revisit the phenomenon of the coastal storm and its propensity to hug the coast. That such behavior of coastal storms is common is underscored in Figure 2.11, which shows the track of two very intense and disruptive coastal storms. They were so famous that they were given individual names. Tracks of these two storms, called the Queen Elizabeth (QE) and President's Day (P) storms, so named because the former storm damaged the ocean liner *Queen Elizabeth* and the latter occurred between the birthdays in February of Presidents Washington and Lincoln. The track of the coastal storm shown in Figure 2.10 clearly favors the large temperature gradient between land and sea. Indeed, this sea temperature gradient is responsible for the wintertime feature, the coastal front.

While the sea temperature gradient—and therefore the surface air temperature gradient over the ocean—is largely fixed by the ocean temperatures (i.e., the location of the Gulf Stream), the land temperature can vary according to the weather pattern. In these cases represented in Figure 2.11, a strong temperature gradient over the land is established alongside the strong (and semipermanent) sea surface temperature gradient, thereby providing a wide corridor of very large temperature contrast, about 25°–30°C (40°–50°F) over a distance of just a few hundred miles (about 500 km).

Let's look more closely at how these storms navigate and how they intensify with the aid of a simple diagram (Figures 2.12a and 2.12b). In the former we have drawn a schematic and highly simplified isotherm patterns resembling a musical stave, upon which are drawn low and high pressure centers with a pair of isobars around each center with arrows to indicate the direction of the wind. Centers of cold- and warm-air advection are labeled as C and W. Lines on the staves are isotherms.

Warm-air advection areas will tend to be associated with convergent surface winds and therefore with ascending motion and falling pressures—and vice versa for cold-air advection. The low will migrate toward the region of maximum pressure falls (to the right of its center), and the high will migrate toward the region of maximum pressure rises. Depending on the strength of the advections, the low and high pressure systems could remain at the same intensity, weaken, or strengthen.

In time, the pattern of isotherms will tend to be move and be deformed by temperature advection and, subsequently, the wind field. This is shown in Figure 2.12b. Several processes are occurring here. First, the areas of convergence

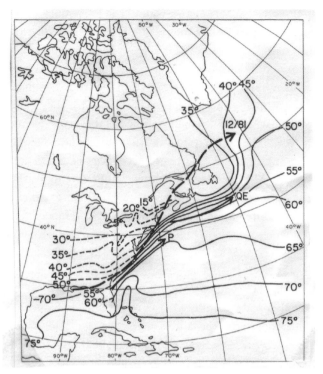

FIGURE 2.11. Representative surface temperatures over land (dashed lines) and over water, the sea surface isotherms (solid isotherms), upon which are shown tracks of three coastal storms: the Queen Elizabeth II storm (QE), the President's Day storm (P), and the coastal storm shown in Figure 2.10 (dashed trajectory labeled 12/81, refers to the date of the storm). (Carlson, *Mid-Latitude Weather Systems*)

and divergence are shifted, respectively, northward and southward by the advecting winds, which are northerly west of the surface low and southerly ahead of it. Second, the isotherms are pushed closer together by the winds, thereby increasing the strength of the temperature gradients (and therefore the intensity of the fronts) and the temperature advections, concentrating them in increasingly narrow zones. Because of this shift in the convergence/divergence pattern, the high pressure center moves eastward but, because of the shift in the advection patterns, it moves eastward with a component of motion toward the south. Similarly, while the low also moves eastward, it develops a component of motion toward the north, following the area of maximum warm-air advection and surface pressure falls. The figure illustrates that fronts also intensify during cyclogenesis in a process called frontogenesis.

Thus, a sharpening of the temperature gradients associated with the frontogenesis process is caused by the ability of the wind pattern to concen-

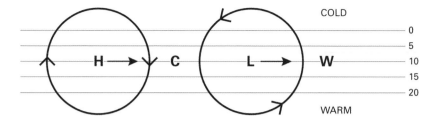

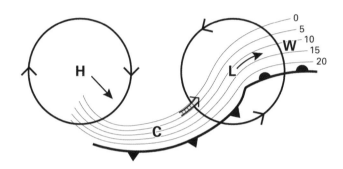

FIGURE 2.12. Schematic illustration of the processes occurring during cyclone development; top: surface isotherms (solid lines labeled in degrees Celsius), the centers of high and low pressure (H and L) with representative wind directions (barbs), and the locations of maximum warm air advection (W) and cold air advection (C). The movement of both the high and low pressure centers is indicated with boldfaced arrows, both pointing toward the right-hand side on the top figure. The bottom figure is the same as the top figure but with the pattern shifted as the result of the effects of temperature advection on the pressure changes. The hatched arrow indicates the location of the jet stream wind maximum.

trate the isotherms. In turn, this causes the temperature advections to become stronger and more sharply defined, resembling the schematic cyclone examples presented in Figures 2.1–2.4. Accordingly, the stronger temperature gradient and (by implication) the stronger winds associated with the deepening low pressure center combine to create an accelerating cycle of development of both the fronts and the low pressure system. For a short time the process is a runaway instability, which becomes checked by rapid distortion of the isotherms and their attendant advection patterns, ultimately leading to a cessation of the advections and the end of cyclogenesis. As noted, this entire process can occur during a period of less than 24 hours.

A further development implied in the figure is the creation of an upper-level jet maximum, which we have referred to as a jet streak. For reasons that lie beyond the scope of this book to fully explain, the jet stream, which is a loose term referring to the belt of maximum westerly winds in the high troposphere, tends to correspond to the region of strongest lower-tropospheric temperature gradient. Because the temperature gradients are larger in the lower troposphere than throughout the rest of the troposphere except near the tropopause, the tropospheric temperature gradients tend to be reflected in the surface isotherm pattern. Accordingly, the low-level temperature gradients tend to mirror the location of the upper jet streak. Thus, the surface temperature pattern provides a crude indicator of where these jet streaks are—warm air to the right of the jet streak and cold air to the left. A clever meteorological analyst could construct a very rough analysis of the upper trough–ridge and jet streak patterns by simply analyzing the surface temperature field.

It was once believed that temperature fronts and jet streaks were a root cause of cyclogenesis. In fact, jet streaks, cyclogenesis, and frontogenesis are interdependent processes. A remarkable aspect of cyclogenesis, as suggested in Figure 2.12, is that the horizontal temperature gradients (fronts), as shown by the spacing of the isotherms in this figure, are enhanced by the cyclogenesis process, leading to an enhancement of the cyclone development. This interdependency and cycle of events reach, simultaneously, to all levels in the troposphere. Since the location of the jet stream and the location of the strongest temperature gradients near the surface are also interrelated, the figure shows that enhancement of the temperature gradient produces a strengthening or formation of a jet streak more or less parallel to the surface isotherms (at right angles to the surface temperature gradient) south and southwest of the low center. Not shown in the figure is a second jet that appears near the downstream ridge where there occurs a strengthening of the horizontal temperature gradient associated with warm-air advection into the downstream ridge north of the warm front.

Thus, it is a mistake to think of fronts or the jet streaks as purely causative factors. The core of the jet stream will often appear aloft in parallel to and on the cold side of strong surface fronts, but some distance from the front, as suggested in Figure 2.12. Cyclogenesis, therefore, involves the creation of a localized jet maximum south and southwest of the low center, more or less parallel to the cold front, which will intensify as the surface cyclone intensifies. The formation of a jet maximum of 55 m/s (about 110 kt) at an elevation of 10 km occurred in our example of a mature cyclone shown in Figure 2.9b.

In this case, the jet streak has formed, not coincidentally, near the region of strongest horizontal temperature gradient imposed by the presence of the ocean and is oriented parallel to the cold front and almost parallel to the coastline. The dry tongue also marks a relatively cloud-free region created by the descent of very dry stratospheric air into this salient.

In summary, the surface and upper-air patterns, the jet stream, and the surface fronts are intimately linked through both the temperature and the winds. It bears repeating that the jet streak is as much a product of the cyclone development as the latter is a product of the former. While it is clear in Figure 2.12 that, empirically, a point at the surface will be relatively warm or relatively cold, depending on the location with respect to the jet stream, neither the surface temperature field nor the jet stream acts independently. Both are closely linked by the physical processes that govern the atmosphere. They are simply two sides of the same coin. As illustrated in Figure 2.9, cyclone intensification accompanies (1) increasingly sharp temperature gradients on the cold side of the fronts, (2) the formation and enhancement of upper jet streaks within the envelopes of the surface temperature gradients, and (3) a migration of the jet streak closer to the surface low in tandem with the appearance of the dry tongue. This is not to say that a single feature, such as a jet streak, cannot be used profitably to diagnose or predict cyclone behavior or development.

Having discussed the larger-scale aspects of coastal storms in this chapter, we will return to this important weather feature with more focus on the smaller-scale features in Chapter 4.

The life cycle of cyclones

Most low pressure systems do not undergo explosive deepening or any intensification. We have seen that one key to the development of cyclones is the presence of a strong horizontal temperature gradient. We will introduce two more related causes later in this chapter.

Cyclogenesis is a more spectacular manifestation of the sideways instability introduced in Chapter 1. While the presence of a horizontal temperature gradient is necessary for midlatitude cyclones to exist and to move about, explosive deepening is favored in certain geographical locations, one of which lies along the eastern coastline of the United States, where strong horizontal temperature gradients exist—at least during the cooler seasons. Not surprisingly, the warm waters of the Gulf Stream constitute a favorite area for cyclone development, particularly when conditions create an even stronger temperature gradient following an intrusion of cold continental air along the coast.

Let's follow the tripartite sequence of cyclogenesis development in Figure 2.13, which begins with the formation of a weak surface low along a frontal zone. The figure shows the area of sustained precipitation (shading), isobars (solid contours), the direction of the vertical motion (upward or downward arrows), and the surface isotherms (dashed contours). Neither the isotherms nor the isobars are labeled, although it should be evident that the pressure increases outward from the center of the low and the temperature decreases with increasing distance of the isotherms from the cold and warm fronts.

The familiar ascent/descent dipole pattern conforms to the warm/cold advections. Following the initial stage, in which a weak low lies within a strong horizontal temperature gradient, the mature, open-wave cyclone and comma-shaped precipitation pattern appears (middle panel). Finally, the storm reaches its maximum intensity and begins to occlude (lower panel). In the last stage, a cold pool of air migrates close to the center of the low, causing the temperature advections around the low to begin to weaken due to the weakening of the temperature advection, as the isotherms and isobars become increasingly parallel to one another, so that advection boxes become larger or vanish completely.

The top figure closely resembles some of our earlier examples of cyclone development: Figure 2.9a corresponds to the upper left-hand map in Figure 2.10 and the top panel of Figure 2.13. Figure 2.9b corresponds roughly to Figures 2.1–2.4 and to the middle panel of Figure 2.13 and the second and third panel in Figure 2.10. The bottom panel of Figure 2.13 is the classic occlusion stage, corresponding to the analysis in Figure 2.8 and the lower right-hand panel in Figure 2.10.

Thus, the end of advections is the end of the storm, although the low itself may be accompanied by rather high winds (isobars close together), if not much sustained precipitation. Patchy shading in the occluded stage signifies that the precipitation pattern has begun to break up into widely scattered showers. In the last stages of cyclogenesis (the weakening stage), closed isotherms corresponding to a cold pool of air move over the low center. This feature exists only partly as the result of cold temperature advections. In fact, it is a consequence of an anomalous situation in which the cold air behind the low had been ascending so rapidly that it cooled even in the absence of cold advection.

As the advections weaken, strong ascent actually cools the air even where cold advection or weak warm advection is occurring. This anomaly is referred to earlier in this chapter and constitutes a deviation from the rule of thumb that warm (or cold) advection accompanies a local warming

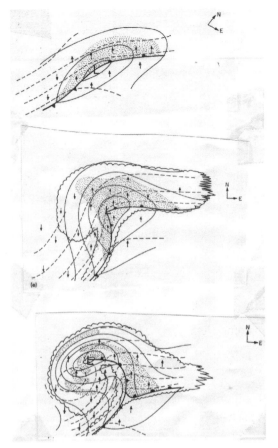

FIGURE 2.13. Three stages in cyclone and frontal evolution. Solid lines are sea-level isobars, dashed lines are the surface isotherms, shading shows the area of sustained precipitation, the scalloped border represents the outline of overcast cloud cover, and small arrows denote the direction of the vertical motion (up or down). From top to bottom are the initial stage; middle; the mature open wave; and the occluding frontal cyclone. Note that the advection boxes become smaller and more numerous as the storm intensifies. (Carlson, *Mid-Latitude Weather Systems*)

(cooling). Instead, the effect of ascent and descent dominate the effect of advection on the temperature field. Rising air cools the column, eventually creating a pool of cold (yet ascending) air. The cold pool gradually drifts over the center of the low.

The appearance of a cold pool near and, eventually, over the center of the occluding low is therefore a consequence of rising air within a limited region of weak cold-air advection. In this anomalous situation, a cold pool of air is left aloft near the surface low center. Not surprisingly, since the pool of cold air aloft corresponds to a steep decrease in temperature with height in the lower troposphere, the cold pool coincides with convective rain or snow showers.

As the storm intensifies the westward displacement between the surface feature and upper trough decreases. The westward displacement of the upper trough (the upper vortex) from the surface low pressure center—some may refer to this displacement as a lag—is necessary for significant temperature advections to take place in the lower troposphere. When advections cease, both the cold pool and the upper trough or vortex drift over the surface low; as a result, the winds, isobars, and isotherms at all levels tend to become parallel to one another. This disappearance of the lag between upper and lower systems is not the result of a simple mechanical process in which the upper wave moves faster eastward than the surface low, the former overtaking the latter. Instead, the elimination of the phase lag between upper and lower systems is a necessary dynamical process of cyclogenesis simultaneously occurring at all levels. Whereas this process of cyclone development tends to reduce or eliminate the temperature advections near the low center, strong temperature gradients still exist elsewhere once end stage of occlusion is reached. In so doing, the atmosphere is imposing its own check on further development.

Where does the energy for cyclogenesis come from?

The necessary condition for what we have called sideways ascent and sideways instability, the temperature gradient, is by far the strongest in the region where the jet stream is strongest. Since the jet stream and the maximum tropospheric (and therefore the surface temperature) gradients are related, the surface lows, which tend to follow surface temperature gradients, will appear to be steered by the winds aloft, though this steering is rather different from that of a sailboat racing with the wind. However, as we have seen, the wind fields associated with lows and highs, especially developing lows, are able to distort the temperature pattern (as in Figure 2.12), thereby distorting the upper flow, which is intimately related to the configuration of the temperature pattern throughout the troposphere. Cyclogenesis is therefore the result of a very transient instability, which is self-limiting as the result of the distortion of the temperature and pressure fields, evolving so as to terminate the temperature advections.

Fundamentally, what is happening when cyclogenesis occurs? When I take my notebook and drop it on the floor, the gravitational force acting on the notebook is a kind of potential for motion until I release it. Subsequently, the gravitational potential is converted to energy expressed by the motion of the falling notebook referred to as kinetic energy; that energy of motion is in turn converted to sound and air vibrations when it hits the floor.

Returning to the analogy with the tightrope walker in Chapter 1, imagine that the tightrope walker completely loses balance and falls from the rope. Prudently, a safety net situated just a meter or two below the rope prevents a more serious fall. In this case, only a portion of the potential energy that can be derived from gravity is released as motion during the fall, since the tightrope walker does not fall all the way to the floor. Similarly, as the cyclone develops, only a small portion of the potential energy wrapped up in the temperature field can be converted to the kinetic energy of motion. When the advections cease near the end of the storm's active lifetime, temperature gradients can remain quite strong, yet only a small amount of that potential energy wrapped up in the temperature field is available for storm development. Indeed, as the storm dies, some of that kinetic energy of motion is converted back to the potential energy of the temperature field.

As the isotherms and therefore the temperature gradient become increasingly distorted, the energy available for conversion to motion is exhausted, just as it is when the notebook hits the floor. In so doing, the advections ultimately weaken, resulting in the temperature pattern losing its ability to generate vertical and horizontal motion of the air. Cyclogenesis continues until all the energy available for cyclone development in the temperature gradients has been extracted. The cyclone continues to exist but thereafter starts to wear itself down, eventually disappearing by being absorbed in the larger-scale pattern. The temperature pattern can be seen as a potential for cyclogenesis, whose energy can be tapped by the advecting winds under the appropriate conditions.

Static stability revisited: Why are lows so much smaller and more intense than highs?

We now return to the question of why cyclones are generally smaller and much more intense than anticyclones, and why some lows (but never high pressure systems) undergo explosive behavior and others do not. Clearly, one cause of cyclone development is the strong horizontal temperature gradients along the land–sea boundary. This alone, however, does not explain why anticyclogenesis does not occur very rapidly and why highs are so much larger than lows. As we have indicated earlier, two other mechanisms besides temperature advection govern cyclogenesis.

In order to explain this asymmetry between highs and lows, we must return to the idea of static stability discussed in Chapter 1 and again earlier in this chapter. In Chapter 1, we used the analogy of Jerry holding the surfboard over a submerged soccer ball to represent the essential static stability of the atmosphere. Recall that we discussed two forms of instability: (1) the instability manifested in convection, which is to say cumulus-type clouds, such as shown in Figure 1.8, and (2) instability associated with sideways ascent, which is a much wider and (usually) more gentle instability driven by horizontal temperature gradients and manifested in the sky by the stratiform clouds shown in Figure 1.9. We used the analogy of sideways (or slant) ascent and descent as responding to the inclination of the surfboard: the greater the angle the more rapid the ascent of the soccer ball.

The steepness of the surfboard angle with the horizontal is analogous to the magnitude of the horizontal temperature gradient. In the examples presented in this chapter, rapidly deepening cyclones (a manifestation of the horizontal instability associated with slantwise ascent) occur in regions of strong horizontal temperature gradients where the winds are able to produce strong warm- and cold-air advections. It bears repeating that such development essentially takes energy out of the temperature pattern and converts it to energy of motion whereby the low is surrounded by a field of closely packed isobars. The process, however, is self-limiting as the pressure and temperature pattern become so distorted that the advections quickly cease, thereby limiting the intensity of the storm.

But this is not the whole story. Theoretical analysis, which lies beyond the scope of this book, shows that the intensity and size of cyclones are related to the static stability of the atmosphere—in other words, related to the average large-scale lapse rate in the vicinity of the cyclone. Larger lapse rates indicate a smaller and more intense storm. Less stable atmospheres favor small, intense cyclones. Highly stable atmospheres either do not favor cyclone growth at all or allow only for the growth of large and weak cyclones.

That the atmosphere is mostly stable with respect to cumulus convection but not to sideways ascent is clearly evident by the presence of layer clouds such as those shown in Figure 1.9. The presence of a statically stable atmosphere is analogous to the surfboard preventing the vertical ascent due to convection, while allowing sideways motion to take place. As noted, static stability is measured by the difference in temperature (the lapse rate) between two elevations: say, the surface and the middle troposphere. Just for the sake of argument, let's take the midtropospheric reference level to be at 8 km. The smaller the decrease in temperature between the surface and 8 km, the larger the static stability—and vice versa with respect to smaller static stability.

Figure 2.7a shows that the most statically stable of the three soundings is the one located behind the cold front, while the least statically stable sounding is the one in the warm sector, the latter having the largest decrease in temperature.

In view of the relationship between static stability and cyclone intensity, cyclogenesis will therefore develop more rapidly and the cyclone become more intense where the surface to 8-km temperature difference is larger, which is to say also less convectively stable. In effect, static stability acts to restrain cyclone development, but the restraint is weaker where the lapse rate is greater. A large lapse rate between the surface and middle levels can occur where the surface air is warm, as in the warm sector or over the ocean, and/ or where the midlevel temperatures are particularly low, as within a trough. Since the horizontal temperature gradient at middle levels is much weaker than at the surface, the primary factor governing the static stability is usually reflected in the surface temperature.

Accordingly, the warmest surface temperatures at midlatitudes during the cooler seasons are found over the ocean, particularly the Gulf of Mexico and along the Atlantic coast over the Gulf Stream. Aside from temperature advection, reduced static stability over the ocean (associated with a steep lapse rate between the warm surface and the middle troposphere) constitutes a second and very cogent reason why cyclone intensification favors locations along the coast or over the Gulf Stream. Static stability is much lower over the Gulf Stream than over the interior of the continent.

Conversely, however, the static stability within high pressure systems, where the air is descending, diverging near the surface, and relatively cold, is considerably greater than in low pressure areas; evidence for this statement can be seen in the temperature sounding located west of the cold front in Figure 2.7a. Consequently, the sounding in the vicinity of high pressure systems more closely resembles the sounding representative of a location well behind the cold front than the one in the warm sector. In short, the atmosphere is too stable to allow significant intensification to occur within high pressure systems, and the presence of high static stability favors larger weather systems, typical of anticyclones. Alternately stated, static stability imposes its own form of a strong surfboard within high pressure systems, but the surfboard is weaker and more porous in areas of low static stability.

Besides a strong temperature gradient and low static stability, a third factor is important in cyclone development. As we discussed in Chapter 1 and earlier in this chapter, as air rises it cools at a rate of about 1°C per 100 m (10°C per kilometer), a process that can result in condensation once the air reaches its saturation temperature. When condensation occurs, heat of condensation is liberated, thereby warming the air. Release of latent heat of fusion (rather than the latent heat of condensation) is the reason liquid in a glass of whiskey on the rocks will remain at freezing (0°C) as long as the ice cubes in the drink continue to melt, even if the ice cubes have temperatures well below freezing. It is also the reason growers will spray their orange crops with water on nights when the temperature is anticipated to go below freezing. Freezing releases heat of fusion (water turning to ice), thereby keeping the wetted leaves at near freezing rather than letting them freeze.

Since most of the condensation takes place over the lowest few kilometers, the condensation warming due to lifting serves to decrease the static stability of the air column, by warming it by release of the latent heat of condensation, primarily at lower levels. This is again the static stability argument, but one that depends on the condensation process. When storms begin to mature and the vertical motion creates widespread and heavy precipitation, the widespread release of this latent heat of condensation further augments the cyclogenesis process. Condensation warming is, of course, greatest where the air near the surface is especially moist, as will be the case over the ocean where all three factors—advection, lower static stability, and condensation in a moist environment—are present.

Lows, therefore, are more intense and smaller than highs. Air ascends in cyclones much more rapidly than it descends in anticyclones, so that the areas of the latter must be much larger and the descent more gentle than in the areas of ascent. Highs, therefore, correspond to regions of gentle descending motion and divergence at the surface.

Geographical areas favorable for cyclogenesis

Warming of the air column by condensation—and its concomitant effect on static stability—is most important where the lower layers are moist and therefore a ready provider of water vapor for condensation. Not surprisingly, oceanic areas, especially warm ocean surfaces, are favored locations for abundant condensation. Over the ocean, the double-edged sword of warm surface air and abundant low-level water vapor are potent factors in cyclone intensification and in explosive cyclogenesis.

Not surprisingly, these three factors are not present everywhere, so that explosive cyclogenesis over the interior of the continent, while not actually rare, occurs mainly during certain seasons and in certain geographical locations, notably during the cool seasons and over the ocean in coastal areas. That explosive cyclogenesis favors the coastal areas, rather than midocean, demonstrates the overriding importance of temperature gradients and temperature advection. Moreover, the absence of explosive cyclogenesis during summer, when strong temperature advections are lacking, even though the surface is warm and the air moist, further demonstrates that temperature advection is still the prime cause of explosive cyclogenesis. While the latter can take place over the land, it is not a frequent occurrence.

In summary, during the warmer seasons, cyclogenesis and, of course, explosive cyclogenesis are much more infrequent than in the colder seasons. Principal reasons for this seasonal dependence are that the temperature gradients (and therefore the attendant temperature advections) over land, and between land and ocean, are much weaker than in the colder seasons. Static stability as defined in this chapter is actually lower in summer, at least over land, but in the absence of strong temperature advection the cyclogenesis mechanism is quite weak despite the stability factor.

Cyclogenesis in mid-continent
Explosive cyclogenesis can occur over the interior of a continent of North America, although it occurs less frequently than along the East Coast. Occasionally intense cyclones develop in association with a strong cold-air outbreak from Canada west of an existing upper-level trough, enhancing the cold-air advection into the vicinity of the trough axis. This outbreak occurs more or less concurrently with enhanced cold and warm advections, an increase in the amplitude of the trough and also that of the downstream ridge, and the development of an intensifying frontal system and a low pressure center. Subsequent events closely follow our description of coastal cyclogenesis.

These outbreaks of exceptionally cold air often accompany widespread and heavy snow over the Great Plains, spreading both north and west of the low pressure center and stranding transportation as well as creating dangerous conditions for wildlife and domestic cattle. While the low moves with a northeasterly trajectory in concert with the upper flow, the Ohio River valley and northeastern states may lie in the warm sector of the disturbance, so that these regions experience elevated temperatures that can last for a few days until the cold front passes. This situation is the well-known *January thaw* that occurs in the northeastern part of the United States during midwinter.

Surface friction and convergence
It is worth mentioning the effect of friction on cloudiness and precipitation. We pointed out earlier that friction perturbs the balance between Coriolis and pressure, and that a balance between those two forces and the frictional

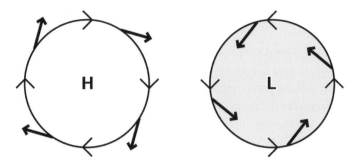

FIGURE 2.14. Schematic representation of a high and low (H and L) pressure system, showing the surface geostrophic flow, clockwise around the high and counterclockwise around the low (thin arrows), and the wind subjected to frictional forces (heavy arrows), outward from the high and inward toward the low. Shading of the low represents the formation of low cloud induced by the frictional convergence.

force requires that the air does not blow parallel to the isobars but with a component of motion across the isobars toward lower pressure. The stronger the winds (the closer the isobars are spaced), the larger the cross-isobaric component. Thus friction will generally cause convergence and upward vertical motion in areas of cyclonically curved isobars within a shallow layer near the surface, the layer affected by surface friction. Even when the large-scale vertical motion is downward, frictional convergence near the surface may be able to produce some upward motion in a shallow layer of cloud. This effect is illustrated in Figure 2.14, which depicts highly idealized high and low pressure systems characterized by cross-isobaric flow outward from the high and inward toward the low, resulting in an often circular-shaped cloud formation (denoted by shading) in the region of cyclonically curved isobars. Figure 2.8 illustrates the circular nature of the cloud mass and its relationship to the curved isobars west of the surface low.

Figure 2.14 is, of course, a gross simplification of the actual distribution of frictional convergence and divergence patterns around highs and lows. Since the cross-isobaric flow, such as depicted here, depends on the strength of the winds, the distribution of frictionally induced stratiform cloud is not evenly distributed about a low pressure center. As illustrated in Figure 2.1, isobar spacing, a measure of the wind speed, is closer together west of the low pressure center than to its east. Accordingly, we are not aware of the frictionally induced stratocumulus layers as the storm approaches but only after it has passed. These low-level clouds are therefore mostly visible from satellite imagery in the region of the comma head west of the storm center.

In the next chapter we will start to interpret the sky: specifically, how the various cloud forms relate to the standard cyclone model and to the processes described in the first two chapters. Of course, we will also consider what the clouds signify in terms of future weather. Chapter 3 will therefore begin by describing each of the different cloud types listed in Table 1 at the end of Chapter 1.

CHAPTER 3
CLOUDS AND HOW TO READ THE SKY

What are clouds?

They speak, but without words. Each day, an array of clouds can pass overhead and yet not everyone takes the time to listen to the message they convey. The clouds in the sky offer keen insights into the state of the atmosphere. They tell us about the atmospheric streams from which they formed. They whisper to us secrets of tomorrow's weather, but we need to learn the language of clouds to understand the story and the wonder that they tell.

Clouds are simply visible manifestations of liquid water in the atmosphere. Moisture as a gas (water vapor) is invisible yet present in minute quantities even in the world's driest places, such as the Sahara Desert and Antarctica. When condensation occurs due to cooling of a rising air parcel (and sometimes even when the water vapor is near saturation), tiny droplets can form on even smaller particles suspended in the air. Substances in evaporated ocean spray compose the majority of these condensation nuclei. Minute particles of sea salt are water attractors known as hygroscopic (water attracting) nuclei, but airborne dust particles from a variety of land surfaces also act as hygroscopic nuclei. Together, these account for the vast majority of atmospheric solid particles (called by the fancy name of lithometeors) that cause cloud droplets to form. Surprisingly, in very clean air lacking in condensation nuclei, liquid cloud drops form at temperatures far below zero

FIGURE 3.1. Swelling cumulus towers viewed from an aircraft.

(both Fahrenheit and Celsius). The relative purity of the condensed water and surface tension forces on the tiny droplets are factors that keep the miniscule droplets from freezing. However, as readings reach −40°C/−40°F, water vapor can exist in a supercooled state but will immediately freeze onto freezing nuclei and form tiny ice crystals. This process known as deposition occurs only in frigid environments, such as high in the atmosphere or closer to the surface in polar regions. Freezing nuclei are far fewer in the atmosphere, so freezing onto solid nuclei does not happen as routinely as does condensation. Supercooled water frequently exists at temperatures not far below freezing, causing grave problems in icing for aircraft.

Expanding on our discussion in Chapter 1, because of random collisions between droplets—a necessary process for the growth of small droplets to the size of raindrops—the size of cloud droplets varies internally. So also does the depth of the cloud, which plays an important role not only in how long droplets can reside in a cloud and therefore continue to grow but also in how much light can pass through to the ground, an indicator of how dense the population of droplets is within a cloud. Obviously, the larger and more numerous the drops and the deeper/taller the cloud, the less the light that penetrates the cloud, and it appears darker. So, wispy, thin ice crystal clouds (smaller in size than droplets) will allow most of the sunlight to pass through, and the clouds will look white as they evenly scatter sunlight. Puffy water droplet clouds will appear variegated with bright white lobes along with shadowy dark corners. The lighting around water droplet clouds depends as much on the sun angle and other clouds nearby as it does on the geometry of cloud shape.

As we pointed out in Chapter 1, water droplets need to reside for a period of time before they can grow large enough to fall out as raindrops. Clouds associated with sideways ascent are long lived, and so are able to produce raindrops even when they are relatively thin. Clouds associated with convection, such as that shown in Figure 3.1, need to exist for more than a few minutes and grow to some considerable depth, at least at midlatitudes, before rain can occur. Figure 3.1 shows a swelling cumulus cloud that, though beginning to acquire some considerable depth, may not yet be producing raindrops.

Cirrus clouds (ci): the cloud whisperer and wanderer

The streaks of white that resemble an Impressionist's paint brush trail across the sky are cirrus clouds, denoted in Table 1.1 as ci. Named *cirrus*, from Latin for "curl of hair," these ice crystal clouds are the highest that can occupy the troposphere; they are typically found at altitudes of 6–12 km (and sometimes much higher). Cirrus clouds are formed by several different mechanisms. These clouds tend to be travelers, frequently originating as the outflow of tall thunderstorms hundreds or even thousands of kilometers away. Often, the origin of this cirrus is thunderstorms occurring in the tropics. The wispy trails are not infrequently entrained into the southerly flow east of deep troughs in the westerly midlatitude flow, appearing atop the comma-cloud pattern described in Chapter 2.

Some cirrus appear as a result of a jet engine's exhaust of water vapor. They are called condensation trails or "contrails"; under certain temperature and water vapor conditions high in the atmosphere, they can grow and spread out, covering the sky for thousands of square kilometers (Figure 3.2, right side). Contrails tend to form when the atmosphere at those levels is close to saturation with respect to ice (ice saturation occurs at less than 100% relative humidity with respect to water). Thus, contrails often are accompanied by or precede the more prosaic cirrus clouds.

Some cirrus clouds are produced by the uplift of air over tall mountain ranges such as the Rockies or the Andes. Other cirrus are the product of a gentler upglide of air over cold or warm fronts, the sideway lift spoken about in the previous two chapters (see Figure 2.6a). Whatever the means of their origin, cirrus speak very softly, even a whisper of the motions of the atmosphere near the jet stream. The way they cross the sky tells us of the speed and direction of the fast air currents overhead, and occasionally, when sunlight illuminates them in a certain way, they tell us of changes that will soon take place in our weather.

Cirrus clouds are carried along by the jet stream and are almost always oriented along the jet. This means that when one traces the direction of the cloud filaments, they are observed to be aligned with the wind direction at that level. For example, the orientation of the cirrus clouds in Figure 3.2 (left side) suggests a wind direction from (or toward) the upper left corner of the figure toward (or from) the lower right corner. In a later chapter, we

FIGURE 3.2. (left) Cirrus growing from condensation; (right) Cirrus from jet condensation trails.

will explore how the differences in the wind direction between the surface and cirrus cloud level can tell us something about whether the atmosphere is warming or cooling.

Let us return to Figure 2.3, the typical cloud and precipitation pattern surrounding an open-wave stage in cyclogenesis, the comma cloud. In referring to the comma-cloud pattern, we include a more detailed version of Figure 2.3 as Figure 3.7. Descriptions of the various cloud forms are placed within the context of this figure, the classic comma-cloud model, labeling the clouds with their abbreviations, as shown in Table 1.1 of Chapter 1. For example, cirrus clouds are abbreviated by the letters ci. It is to be understood that the observation is being made at the surface and that the symbol in this figure indicates more or less the dominant sky cover of that type of cloud visible to the observer on the ground; it is also understood that the sky may be partially covered by that type of cloud, the remainder being either clear sky or another type of cloud. When two types of cloud are visible from the surface, we will indicate that situation by a slash (e.g., ci/cs). Higher clouds can, of course, be obscured by lower cloud layers, so being unable to observe the presence of cirrus clouds, for example, does not preclude their presence if they are obscured by lower layers of altocumulus, altostratus or stratus.

Cirrus clouds exist in a many forms, there being in the nomenclature a handful of varieties. The most familiar are cirrus uncinus (or, curly hooks/ mares' tails), cirrus aviaticus (or contrails; see pictures above), and cirrus spissatus (or remnant anvil cirrus from a thunderstorm; see Figure 3.3). However, we will generally confine our discussion to the types of clouds represented in Table 1.1.

Referring to Figure 3.7, the first sign of an approaching storm is found at the eastern end of the cloud system. Many times, cirrus clouds, first appearing in the western sky, are the first sign of a change in the weather pattern, as

FIGURE 3.3. Cirrus spissatus formed from the remnants of an earlier thunderstorm. Notice the virga, or shafts of ice crystals, falling from the cirrus. All of these ice crystals are evaporating in the dry air below the cloud level.

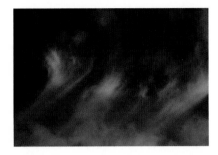

FIGURE 3.4. Cirrus clouds as mares' tails, showing the puffs and their trail of ice crystals.

they herald moisture being lifted far away and carried toward the observer. This is more likely the case in the colder half of the year when storm systems are larger, move faster, and are better organized. Yet even in the summer, a shield of cirrus clouds often signals nearby weather action, although summer cirrus may also be a herald of distant thunderstorms, as mentioned above. Cirrus clouds have been nicknamed mares' tails in keeping with their original name, curl of hair. An old folk saying speaks toward the progression of clouds preceding a typical storm: "Mares' tails (cirrus) and mackerel scales (altocumulus) make tall ships lower their sails." That is, when cirrus clouds are followed by a lower deck of water droplet clouds, a midlatitude cyclone is likely on its way. Wind-powered ships would do well to lower their main sails so that they can navigate in sharply shifting winds, and to avoid having their sails tear as winds rise to gale force.

The physical origin of mares' tails is that the tail is actually a trail of ice crystals falling from a cirrus puff and bending with the wind direction at that level. The puff is maintained by condensation, which continuously generates the source of the ice crystals, while those in the tail eventually evaporate at lower levels. A good example of mares' tails showing the puff and trail structure is shown in Figure 3.4. Evaporation of ice crystals takes place very slowly, so that cirrus can maintain themselves for many days after they depart the saturated environment of a cyclone.

FIGURE 3.5. (left) A halo around moonlight through cirrus; (right) a mock sun (bright spot above chimney) through cirrostratus.

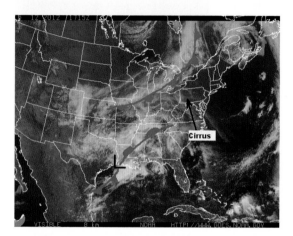

FIGURE 3.6. A visible satellite image of the United States displaying a shield of cirrus preceding a cold front and a weak trailing surface low in the western Gulf of Mexico. The cirrus is just reaching Pennsylvania at the head of the arrow. Courtesy of NOAA/NESDIS.

Since cirrus clouds comprise ice crystals, they act as a lens for sunlight, bending it through the tiny prism and causing a number of optical effects. The popular saying "Halo around the sun or moon means that rain or snow will come soon" has a ring of truth to it in a similar way to the previous saying about mares' tails preceding a storm system. The size, shape, and orientation with respect to the ground of the ice crystals in cirrus clouds can produce a plethora of refraction phenomena ranging from haloes and arcs to pillars and mock suns.

The presence of a halo (Figure 3.5), caused by sun or moonlight being bent at a 23° angle while passing through an array of randomly shaped and sized ice crystals, is the most common optical effect, and about 70% of the time a halo is followed by precipitation within 24–36 hours. (In summer these numbers may have to be doubled.) This foretelling of rain or snow is easy to understand when one considers that cirrus clouds are usually along the advancing perimeter of a well-organized midlatitude cyclone whose

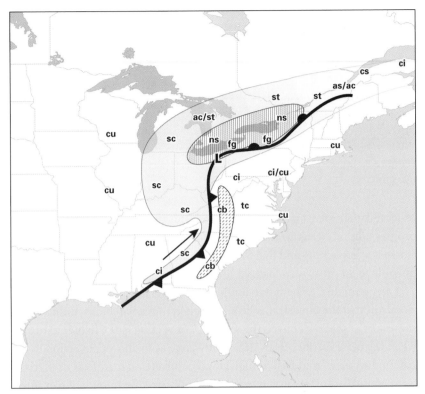

FIGURE 3.7. Comma-cloud pattern associated with a mature open-wave cyclone. This figure is essentially the same as Figure 2.3, except that cloud symbols referred to in the text and abbreviated in Table 1.1 are superimposed on the picture. Symbols refer only to clouds observed from the ground and, as such, may mask multiple layers of cloud above. Where two cloud types are simultaneously observed from the surface both are indicated and separated by a slash. Two symbols not listed in Table 1.1 are: tc (towering cumulus) and fg (fog).

passage can take a few days (see map; Figure 3.7). Figure 3.6 shows just the gossamer wisps of cirrus well north of the frontal system located along the Gulf of Mexico coastline. Since atmospheric motion is usually from west to east, the Ohio River valley and Mid-Atlantic region, covered by cirrus (with a halo possible), would be in the direct path of inclement weather.

Cirrostratus (cs) and cirrocumulus (cc): the harbinger

As the storm moves closer to the observer, cirrus clouds thicken from mares tails to a more general sheetlike pattern, illustrated in the photo on the right in Figure 3.5. This cloud layer may appear thin from a distant surface observation point, but it may occupy a kilometer or so in depth. When halos are

FIGURE 3.8. (left) Partial cover of the sun by cirrocumulus; (right) cirrocumulus covering the entire sky.

observed, they are usually associated with this type of cirrus cloud. A steady progression of cirrus to full cirrostratus cover is further evidence that a storm is approaching and that the cyclone and frontal system have matured, possibly to an open-wave pattern, evocative of Figures 2.1–2.4.

Cirrocumulus, as its name implies, has some attributes of the cumulus clouds in that the cirrostratus layer is broken into small blobs, indicative of some sort of shallow convective activity. Convection (see Chapter 2) occurs when the lapse rate becomes sufficiently large as to allow convection to occur. In the case of middle and high clouds, convection manifests itself in the form of puffs or turrets atop more or less stably stratified clouds. An example of cirrocumulus is shown in Figure 3.8.

Altocumulus (ac) and altostratus (as)

These very beautiful clouds constitute silent but elegant words of change. While cirrus clouds sometimes indicate a shift in the weather pattern, the progression of dominant cloud types from cirriform clouds to altocumulus (or altostratus) is a more certain sign that the weather is changing, as can be seen from inspection of Figure 3.7, wherein the altocumulus and altostratus are located east of the precipitation area but much closer to the low center than are the cirriform clouds. (Note, however, that a more certain harbinger of storminess is the presence of cirriform cloud layers above the midlevel clouds.) When these so-called buttermilk clouds appear, their colors and patterns often speak louder than the message they carry. Unlike its higher-up counterpart, the cirrus, these mixed clouds (both ice crystals and water droplets) show many different faces both day and night (Figures 3.9 and 3.10). When arrayed in a long sheet with embedded waves, altocumulus undulatus (undulating) can be stunning when the sun is low in the sky and illuminates the underside of the clouds (Figure 3.9 left side).

FIGURE 3.9. (left) Altocumulus undulates as seen at sunset on a late autumn evening; (right) an undulating altocumulus layer, thickening and lowering in the distance.

FIGURE 3.10. (left) Altocumulus at sunrise over Chesapeake Bay on a late December morning; (right) an extensive deck of altocumulus as seen from an aircraft.

Altocumulus typically occupy the atmosphere between 3 and 6 km. As their name implies, these are high (alto) heaps (cumulus), which indicates that there is more than just gentle sideways ascent at this level; there is the ever-present convection. A wedge of warmer, moist air has led to a layer containing numerous small convective cloud elements with downward motions (holes) in between (called interstices). These layered clouds are often referred to as a mackerel sky, because their small flat sheets tend to resemble the scales of a mackerel (Figure 3.10), although they have some cumulus-like properties and provide some keen insight into the thermal dynamics of the middle part of the atmosphere. When the warm, moist layer rises and its vapor begins to condense, often little towers or turrets will rise from the altocumulus layer—a certain sign of a steep (large) lapse rate, signaling convective instability and the potential for deeper convection in the near future.

Altocumulus castellanus (little castles) is the name given to these types of clouds when they appear to be sprouting turrets. Evidence of these unusually shaped clouds, which more closely resemble little towers rather than blobs, indicates that convective activity may accompany more general precipitation,

FIGURE 3.11. Altocumulus castellanus.

FIGURE 3.12. Altocumulus lenticularis as seen at sunset on a summer evening.

although such castellanus clouds also occur over deserts where the temperature decreases very rapidly with height from a heated surface, indicative of a convectively unstable atmosphere, but one so dry as to support only these castellanus clouds. The presence of towers indicates a reduced static stability of the atmosphere. An example of this type of cloud is shown in Figure 3.11.

The name altocumulus (and cirrocumulus) implies that these clouds are subject to some convective instability, at least over a shallow layer containing the clouds. Unless the altocumulus clouds proceed to sprout towers, as previously described, the convection is representative of only a shallow layer. Indeed, while aircraft flying within an altocumulus layer might experience some mild bumpiness, they would encounter smooth air just above these types of clouds.

Following our analogy in Chapter 1, the tilted surf board is somewhat porous, allowing both sideways and vertical ascent to take place simultaneously. In the case of altocumulus or cirrocumulus, however, that convective instability may be limited to a shallow layer containing just the altocumulus. In the absence of these shallow convective layers, the midlayer clouds are altostratus. Both altostratus and altocumulus tend to exist together, the former increasingly dominant as the clouds continue to thicken. When preceding a storm, altocumulus and altostratus tend to exist in conjunction with cirrus layers above.

On other occasions, the puffiness of the altocumulus clouds is damped by dry air aloft leading to distinct up and down motions, manifested as a

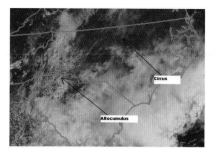 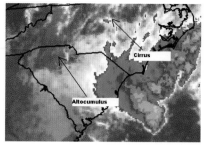

FIGURE 3.13. Geostationary satellite images from the same time and location (mid May over the Southeast United States). (left) A visible satellite image upon which is superimposed radar echoes, yellow corresponding to heaviest rainfall; (right) arrows show locations of altocumulus and cirrus clouds. Courtesy of NOAA/NESDIS.

FIGURE 3.14. Altocumulus with a faint ring of color (iridescence) surrounding the partially obscured disc of the sun.

wavy structure within the cloud layer. When this is enhanced by air flow across a mountain range, the clouds can take on a lens shape that occasionally resembles flying saucers (Figure 3.12). This form of altocumulus cloud is called altocumulus lenticularis.

Waves within altocumulus clouds align parallel to topographic features and usually at right angles (orthogonal) to the wind direction. From above (Figure 3.13)—that is, from the vantage point of a geostationary satellite—the interstices in the altocumulus are too small to distinguish it from cirrus. Though in a visible satellite image (left), these clouds look less fibrous (they contain less ice) and also are notably warmer (lower in the atmosphere and darker) (right) in the infrared images than their first cousin the cirrus cloud.

Altocumulus, because of their water droplet constituents, can cause a diffraction effect known as iridescence. This optical phenomenon appears as a ring of rainbow colors around the sun, or around a portion of the partially obscured sun. Unlike refraction, which bends the light, this effect is caused by many tiny droplets diffusing the sunlight into its full spectrum of colors. This effect is sometimes confused with a halo, as shown in Figure 3.14, but it occupies a much smaller circle closer to the sun and is distinctly more colorful. It is also a visual precursor of an approaching storm. Like the analogous situation for cirrus, thin altostratus or altocumulus can also form a halo around the moon.

The hushed threats of a storm

The presence of cirriform clouds followed by layers of altostratus and altocumulus clouds, highlighted by the way the light from the sun or moon is steadily swallowed up by these clouds, is a nearly sure sign of a storm. Yet, even as altocumulus and altostratus appear and thicken, cirriform clouds can still be seen through spaces in the lower cloud decks. Indeed, the presence of both clouds is a necessary precursor of an approaching storm. As mentioned earlier, altostratus means high (alto) flat (stratus) clouds that are relatively flat and opaque, and generally diffuse enough sunlight so that shadows are not distinguishable. The water content of these clouds is high enough that they appear as a uniform gray color and generally are seen a few hours before it starts to rain or snow. Due to the mixture of ice and water in altostratus, the sun and moon appear watery when their light is able to shine through. The dim lighting, indistinct grayish appearance of the sky, and often a freshening breeze are hushed reminders that a storm is about to begin—a storm that may last for a day or more. As with cirrus sometimes being a precursor of thunderstorms rather than a general rainstorm, the

FIGURE 3.15. Altostratus obscuring sunlight so that no shadows are visible on this winter afternoon.

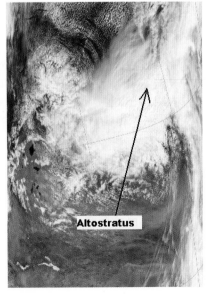

FIGURE 3.16. View at midday of a polar-orbiting satellite, *Aqua*, over central Asia on May 21, 2012. Cirrus clouds can be seen just above the location of the arrow head. Courtesy of NOAA/NESDIS.

residue of a cluster of thunderstorms can be thick enough to leave a deck of altostratus and altocumulus in its wake, although, unlike cirrus, these types of cast-off clouds from thunderstorms signify that the latter are almost invariably close to the observer.

When viewed from above (from a satellite), altostratus is often hidden below a deck of cirrus. Figure 3.16, showing both altostratus and cirrus clouds, is a midday view from the polar-orbiting satellite *Aqua* over central Asia on May 21, 2012. A large midlatitude storm system covers much of eastern Siberia. For the observer viewing these clouds at the surface, it is likely that once altostratus appear it is unlikely that skies would clear anytime soon. Their murmur of impending precipitation will be heard for hours.

The annoying repetitions of the past: Stratus (st)

We can almost breathe in stratus, the lowest of the layered variety. Altostratus and altocumulus clouds customarily thicken to stratus just before the outbreak of precipitation, as illustrated in Figure 3.17. Stratus clouds are easy to identify, but their significance is not always obvious. As often as not, stratus clouds can be the remnants of a storm that has passed by. The key to interpretation is to assess whether the stratus deck evolves from its storm precursors, altocumulus or altostratus and cirriform cloud (as it appears to be in the distance in Figure 3.17), even if it looks solid from horizon to horizon, as in Figures 3.18 and 1.9.

FIGURE 3.17. Altocumulus thickening to stratus in the background.

FIGURE 3.18. Stratus covering the sky across the hilly terrain of central Pennsylvania.

Stratus is part of popular folklore. One saying states, "A summer fog for fair, a winter fog for rain, a fact most anywhere in valley and on plain." This morsel of weather wisdom highlights the different processes that contribute to stratus formation. During the summer, stratus form in a humid environment, typically after it has rained late in the day. As a result, the low clouds the next morning inhibit the sun from heating the ground, which delays the initiation of convective clouds and therefore decreases the risk of a shower—so with a summer fog, the weather will be fair!

When stratus is at ground level, we call it fog. Stratus can make surface travel treacherous and when the cloud is dense enough even air traffic will come to a halt. Ordinary stratus, not usually associated with fog, can be encountered as fog on mountain roads where their elevation is above the cloud base. Some of the more memorable accidents at sea have been in dense fog, and each year major pileups on interstate highways are attributed to dense fog, or stratus. Stratus clouds comprise water droplets, which are supercooled on some winter mornings and can cause a light icy coating on exposed surfaces, particularly bridges and overpasses. Freezing of super-cooled water droplets also poses a problem for aircraft.

Stratus form in a variety of ways other than as harbingers of an approaching storm:

- From moistening of the lowest part of the atmosphere during rain or snow
- From mild, moist air moving over cold ground or chilly water (an example of steam fog is shown in Figure 3.19)
- From condensation of water vapor near the surface on clear, calm nights (see below)
- From somewhat moist air that is forced to rise on sloping terrain, over a shallow layer of cold air, as shown by the symbol for fog just north of the warm front in Figure 3.7, or as the result of slantwise ascent leading to condensation of the air at lower levels

Aside from fog associated with the overrunning of warm moist air from the warm sector (slantwise ascent), the other three types of fog are seldom associated with precipitation. However, in the winter, fog can occur when mild, moist air moves over colder ground, which in the eastern United States occurs with winds from the south. Recall that the circulation around low pressure is counterclockwise so that a south wind portends a storm passing west of the foggy region, leading to warmer weather and showers. Fog is

FIGURE 3.19. A steam fog or stratus forming over a Pennsylvania lake in the late summer.

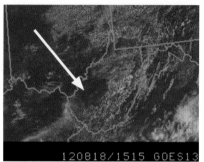

FIGURE 3.20. (left) a large fog bank over West Virginia at 6:45 am EDT on August 18, 2012 is starting to dissipate from the sides (arrow) toward the middle 3.5 hours later (right). Courtesy of NOAA/NESDIS.

often present just on the cold side of the warm front (see Figure 3.7), where the inversion created by the overrunning of warm air is relatively close to the ground (see the warm-front cross section in Figure 2.6b and the warm-front sounding in Figure 2.7a).

Fog not associated with frontal precipitation often burns away during the morning. One peculiarity of fog, as noted on satellite imagery, is that it does not burn away during the morning in a uniform manner but rather shrinks from its boundaries toward the middle. This pattern of fog dissipation is due to heating by the sun just outside the fog boundaries, causing a difference in surface temperature between the heated ground and that underneath the fog. The resulting temperature gradient induces a local circulation in which the heated air ascends just outside the cloud boundary and descends just inside the fog, causing the latter to warm by compression and evaporate. The fog continues to shrink along a progressively drying perimeter until it finally disappears in the center of the previously foggy area. A pair of satellite images showing the progressive burning away of a large fog bank from the sides over a period of about four hours is shown in Figure 3.20.

When in the warm sector of a traveling midlatitude storm during the winter, rain is the favored form of precipitation—just as the folklore indicated, a winter fog augers rain. Occasionally, a significant south–north temperature gradient can exist in the warm sector, resulting in precipitation in this normally quiescent sector; this added complexity may cause the map analyst to draw a second warm front. In wintertime, the progression of cloud heights (whose bases are called cloud ceilings) preceding the passage of a warm front ends with the cloud on the ground, or stratus, just before the warm front passes and the observer becomes located in the warm sector, as suggested in Figure 3.7. This weather lore pertains to the wide-open plains as well as the rolling hills of the Appalachians.

In the heart of the Appalachians, Pennsylvania and surrounding states such as West Virginia have a very peculiar type of hilly terrain that tends to trap cold air in the many valleys between chains of relatively low mountain ridges, which extend about 300–400 m above the valley floors. Warm fronts tend to be drawn as holding back from crossing Pennsylvania in the wintertime, even though the warm air from the warm sector has moved northward at levels close to the tops of the ridges. These situations are often accompanied by freezing rain or sleet in winter, such that rain falling from layers with above-freezing temperatures arrives at the surface where the air is sometimes well below freezing. In this situation, drops almost instantly freeze or have already started to freeze as they penetrate the shallow layer of subfreezing air, producing a glaze of up to a quarter inch of sheet ice on the surface and creating one of the most hazardous weather conditions to be experienced in these areas during wintertime. We will discuss this type of reluctant warm front in the next chapter.

Stratus and nimbostratus

Stratus clouds are flat sheets of cloud typically found between 1 and 3 km above the surface. A good example of a stratus deck is shown in Figure 1.9. Figure 3.21 illustrates that stratus clouds can be mixed in with a variety of other cloud types and are often difficult to identify from satellite; in this case, the stratus layer does not portend an approaching storm. Instead, it lies well behind the surface low and may actually have been formed by the spreading out of a stratocumulus layer, a more fair-weather cloud type. We have emphasized that one way to distinguish stratus as a storm precursor from the more benign stratocumulus is that the former is much more likely to accompany decks of middle and high clouds, whereas the latter would exhibit clear blue sky through openings in the cloud layer.

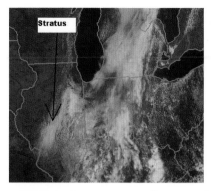 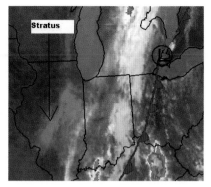

FIGURE 3.21. Geostationary satellite images from the same time and location (mid May over the Midwest of the United States). The left image is a visible view of the clouds over Illinois, while the right image is an infrared view at the same time. Superimposed on this figure are the location of the surface low pressure center and the surface fronts. Arrows show location of stratus clouds. Courtesy of NOAA/NESDIS.

The weak low pressure system and incipient wave development shown in Figure 3.21 failed to develop, probably because of the lack of a strong temperature gradient—and therefore the temperature advection.

It is worth repeating that stratus, when preceded by the progression of cloud types from cirriform to midlevel clouds (altocumulus/altostratus), inevitably signal the immediate onset of precipitation. Just prior to the precipitation reaching the ground, however, the signal that it is about to rain or snow can be seen in hooklike streamers of cloud hanging from the underside of the stratus deck called by the Latin name *virga* or by the more prosaic *precipitation trails*. Virga consist of raindrops that proceed to evaporate into the layer of air beneath the cloud. Evaporation, by requiring an amount of heat to turn liquid into vapor (the opposite process from condensation, referred to earlier, but involving the same amount of heat energy taken from the air by evaporation as that imparted to it by condensation), cools the air beneath the stratus. Gradually, the lower and initially unsaturated air layer becomes sufficiently moist to allow the droplets to reach the surface as rain or snow without first evaporating. In so doing, the subcloud air layer gradually cools. It is not uncommon in such cases to find that the air near the ground, though initially unsaturated and above freezing, cools to a temperature close to or below freezing, resulting in a decrease in surface air temperature that is accompanied by snow or rain changing to snow. This is often the case with warm-front precipitation (precipitation occurring just north of the warm front in Figure 3.7), although these marginal snowfalls often turn (or return)

FIGURE 3.22. Virga beneath a stratus deck.

to freezing rain or rain within a relatively short time. Visually, one often sees banks or rolls of very low cloud almost at surface level hugging the sides of a mountain during these types of rainstorms.

Once precipitation has started, the stratus clouds are referred to as nimbostratus, the prefix *nimbo-* (coming from the Latin for cloud or rainstorm) referring to any rain-producing cloud. Figure 3.22 shows the hook-shaped virga beneath a stratus deck. Some of the rain shafts in the background are reaching the ground.

The billowy boastings of cumulus clouds

They have been described as ice cream castles in the air. Cumulus clouds are from the genus convective cloud, which means that its beginnings are often from air heated by the ground. There are many variety of cumulus with the smaller versions (Figure 3.23) containing almost all water droplets and the larger versions having ice crystals in the upper chambers of the cloud. They can portend both fair and foul weather, depending on their size and structure.

Since these "heaps" (the literal meaning of *cumulus* from Latin) of cloud drops are created by rising thermals induced by sun heating the ground, their presence indicates an unstable atmosphere (at least over the particular layer where they are seen). Recall from our discussion in the previous chapter that the greater the decrease in temperature with height (the larger the lapse rate), the greater the likelihood of convective instability. This instability is further increased by the presence of condensation, which through the release of the latent heat of condensation (the same as the latent heat of evaporation referred to above), warms the air at cloud level (primarily in the lower part of the atmosphere), thereby augmenting the rate of temperature decrease with height above the clouds, while causing saturating air pockets to cool more slowly with height when they are lifted. The process of cumulus cloud formation is an aspect of the buoyancy force previously discussed in Chapters 1 and 2 and in more detail in Chapter 4.

FIGURE 3.23. Fair weather cumulus over the hilly terrain of Pennsylvania at midday in the springtime.

As discussed in the previous chapter, small fair-weather cumulus are formed when thermals rising through a convectively unstable atmospheric boundary layer reach condensation. This process is illustrated in Figure 3.24. Note that the bases of the clouds in Figure 3.23 all appear to be at about the same height above the ground. Cumulus cloud base heights are dictated by the overall difference between temperature and dewpoint in the surface layers: The larger the difference, the lower the relative humidity, the higher the bases. Cumulus cloud bases in dry environments (such as the southwestern part of the United States) tend to be quite elevated, so high that precipitation often completely evaporates before reaching the ground. In the Northeast, however, cumulus bases are typically around 1 km.

The quickening pace of overturning of the lower atmosphere almost always leads to the development of a gusty breeze as the compensating sinking motion in the clear air around the puffy clouds transports stronger winds from above toward the surface. This process, outlined in Figure 3.24, is called mixing. The downward currents between the clouds serve to create clear spaces, a process that tends to order the clouds into rows oriented approximately parallel to the wind direction, while the transport of faster-moving air from above tends to promote gentle breezes at the surface. This process is the reason surface wind speeds often pick up from near-calm conditions during the morning or afternoon, especially when cumulus clouds are present. Mixing brings down faster-moving air from higher levels but also brings up slower-moving air from below.

Fair-weather cumulus clouds, known as cumulus humilis, are as innocuous as they are pretty. These clouds typically extend from their bases, near 1 km to about 2–3 km. Such clouds may extend higher in conditions of weak static stability, although they usually flatten out on top. In hilly terrain, cumuli first form over the heated, sunny sides of ridges during the mid- to late morning (9–11 am local standard time [LST]) as the heated air on the sides of the hills ascends quickly into the surrounding cooler open air. Depend-

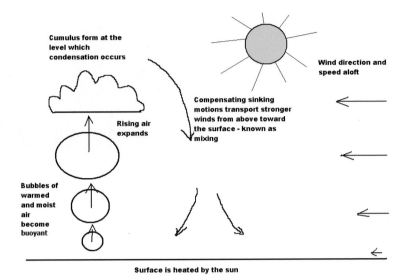

FIGURE 3.24. Schematic illustration of the formation of cumulus clouds showing that the rising thermals in a shallow convectively unstable layer near the ground achieves condenstation at a level dictated by the temperatrue and moisture content of the subcloud layer. Sinking motion occurs around the clouds, forming clear spaces.

ing on the profile of temperature and moisture in the lower atmosphere, the coverage of cumulus clouds usually reaches its maximum in the mid- to late afternoon (2–5 pm). Since cumuli are mainly driven by daytime heating, there are fewer of them at night. Often, the cumulus clouds (and even stratocumulus) will vanish with the setting of the sun. On occasion, when the prevailing winds are especially strong, intermittently strong breezes can persist at night near the surface, causing some cumuli formation, especially over uneven terrain.

What distinguishes fair-weather cumulus clouds from their bigger, taller kin is the environment in which they grow. Fair-weather cumuli tend to exhibit blue sky patches between them and have little vertical development. Figure 3.25 (left side) shows cumulus congestus clouds while the right side of the figure shows some precipitation with cumulus congestus and towering cumulus clouds in the background. The appearance of clear sky with little or no higher stratiform cloud is an indication that the overall weather pattern is that of fair weather, despite the possibility of brief squalls in the vicinity.

Recall from our discussion of altocumulus castellanus that the presence of vertical development in the form of turrets in cumuliform clouds signals a potential for precipitation. Cumulus congestus (e.g., Figure 3.1; see

FIGURE 3.25. Cumulus congestus clouds during an early summer afternoon (left); Cumulus congestus and a precipitating cumuliform cloud (incipient cumulonimbus) with clear sky visible (right).

also Figure 3.25) or towering cumulus clouds form from cumuli but, having greater buoyancy due to the warm moist air rising into the cloud from below its base, experience an additional lift. Preferred areas for congestus development are over mountain ridges, over local hot spots, or lifted over cool air outflow from nearby mature thunderstorms. As congestus clouds continue to grow and produce sizable towers, they may then be referred to as towering cumuli, basically grown-up congestus clouds. Such cumulus clouds have been observed over nuclear reactors and forest fires.

Size and longevity are highly correlated in the cloud world. Fair-weather cumulus clouds tend to be very short lived, lasting less than a few minutes and seldom more than 10 or 15 minutes from formation to disappearance. Most fair-weather cumuli, as observed from the ground, are no longer buoyant, having lost their buoyancy almost as soon as they form due to the ingestion of dry air from the surroundings. Visually, what distinguishes the congestus or towering cumulus clouds is not only the vertical development but also the appearance of what looks like hard, polished surfaces of the towers springing up above the cloud. Ragged edges of cumuli signal the end phase of the cloud, but surfaces that look as though they are made of porcelain are vigorously buoyant, indicative of a deep layer of convective activity. Such towers can be seen at the top of the congestus cloud on the left side of Figure 3.25 and also are evident in Figures 3.1 and 1.7. Congestus and towering cumulus clouds often herald the formation of a thunderstorm.

What makes a cumulus cloud become a congestus?

As their name implies, these clouds are a cluster of cumuli drawn together by more potent rising air currents. There are a number of factors that can turn innocuous cumuli into a hefty cumulus congestus:

- *Changes in the profile of temperature and moisture in the lower atmosphere (cooling aloft and/or warming at the surface).* Recall from Chapter 2 that the greater the lapse rate—the decrease in temperature with height, such as measured between the surface and middle levels, say 8 km—the more convectively unstable the air. Convective development leading to a thunderstorm can occur if the air aloft is relatively cold or the air near the surface is relatively warm and moist. The latter is necessary for severe thunderstorms because condensation provides an added vertical boost to the cumulus clouds.
- *Converging low-level winds that occur ahead of weather fronts.* Given the favorable conditions listed in the previous statement, convergence will provide a trigger for convection. Figure 3.7 shows a line of thunderstorms (cb) preceded by towering cumulus (tc) clouds just east of the cold front and more or less parallel to the front. This location is favored by the fact that air moving up from the south in the warm sector is warmest and most moist in a relatively narrow tongue just east of the cold front. The line of thunderstorms depicted in Figure 3.7 is called a squall line.
- *Forced ascent over rough terrain.* In the absence of an organized squall line, terrain features play a central role in determining where thunderstorms will occur.
- *Localized differences in surface temperature and moisture.* Along a shoreline or on the sunny side of a mountain ridge or over a local hot spot such as a city.

While brief showers can fall from cumulus congestus clouds (many of these clouds do not precipitate), they can also grow into the majestic thunderheads known as cumulonimbus. These clouds need to reach altitudes of at least between 20,000 and 24,000 ft tall (7–8 km) in order for a sufficient separation of charge to occur on the droplets and ice crystals for lightning to be seen. Some cumulonimbus clouds reach altitudes of more than 50,000 ft (16 km). The billowing of cumulus clouds may foretell even greater boastings. Examples of cumulonimbus clouds are shown in Figures 3.26 and 3.27.

Generally, however, severe events such as heavy rain, hail, or tornadoes do not develop until the cumulus clouds begin to spread out laterally at the top in the form of a dense layer of ice crystal clouds (cirrus), the anvil. (Recall our discussion that some cirrus clouds originate from thunderstorms, often at great distances from the observer.) Anvils, such as those shown in Figures 3.26 and 3.27, are formed when the buoyant towers reach the tropopause at elevations of 12 km or more above the ground.

FIGURE 3.26. Cumulonimbus clouds starting to produce an anvil at the top.

FIGURE 3.27. Another view of a cumulonimbus cloud with anvil, illuminated at the top by the setting sun.

The anvil top is where the ascending air currents encounter the tropopause, which, as shown in Figure 2.7a, is the part of the atmosphere where the temperature ceases to decrease with height and may increase somewhat. The tropopause temperature inversion acts as a brake on the vertical development of the cloud. The stable temperature cap at the tropopause is illustrated by the sounding diagrams shown in Figure 2.7a. The result is that moisture associated with the cumulonimbus current is sheared off and carried along by the winds at tropopause level (very close to where the maximum winds in the jet stream are located).

While the tops of ordinary thunderstorms, such as found over the northeastern United States, seldom reach elevations much above 12–15 km, it is not uncommon, especially in the Midwest, for vigorous cloud towers to penetrate the tropopause up to heights of 15–18 km before collapsing back to the level of the tropopause. (This is likened to Jerry's soccer ball mentioned in Chapter 1, having been released from below the surface of the water loses its buoyancy once it has risen above the surface of the water and thereafter falls back to the surface.) Such clouds as those that penetrate the tropopause are especially violent.

What sets the organized thunderstorm apart from the one that pops up, growls for a while, and then disappears is its organization, which allows the rain shaft and the ascending air into the base of the cumulonimbus cloud to

FIGURE 3.28. Mammatus clouds on the underside of a deck of low clouds at the advancing edge of a thunderstorm.

avoid interfering with each other, just as separate up and down escalators facilitate the movement of people. The updraft and the downdraft in these storms are located in separate parts of the cloud. This is the structure of the most severe of all the thunderstorms. Its organization is dependent on having a large vertical wind shear between the surface winds and those at upper levels. The type of atmospheric structure favoring these very severe thunderstorms, which can bring large hail, tornadoes, and damaging winds, is principally due to three factors: (1) the presence of warm and moist air near the surface, with dewpoints exceeding 22°C; (2) dry air in the middle troposphere; and (3) strong vertical wind shear (invariably a strong upper jet streak).

An interesting aspect of the cloud structure associated with thunderstorms is the dark, roiling deck of mammatus clouds that precedes the approach of the rain shaft. Mammatus clouds (Figure 3.28) are literally upside-down cumulus clouds whose little towers, the mammata, appear on the underside of the deck of low clouds. Unlike regular cumuliform clouds, these mammata are descending due to negative buoyancy created by the evaporative cooling of the water within the cloud.

Flamboyant and boisterous shouts

Literally speaking, cumulonimbus are the only clouds that make noise. And what a boom they can produce! Cumulonimbus are thunderstorm clouds, and they virtually always produce precipitation (nimbus for dark gray, rain-bearing cloud). Cumulonimbus clouds, also referred to informally by meteorologists as cbs (CEE BEES), are the tallest of the convective clouds. Because of their vertical size, the upward and downward motions within the cloud are very strong and can exceed 25 m/s, speeds that could easily tear the wing off an airplane or cause it to lose altitude or stall. It is not surprising, therefore, that pilots diligently avoid these types of clouds.

Cumulonimbus can be isolated towers or grow in clusters. They tend to congregate along lifting surfaces, such as fronts, convergence lines, or terrain

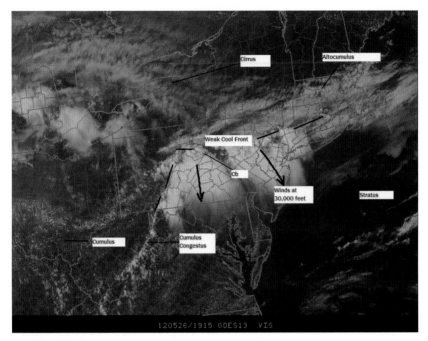

FIGURE 3.29. Satellite image taken over the northeastern United States in late May 2012, showing a variety of clouds. Most relevant to the current discussion is cumulonimbus (cb) embedded in a mass of clouds just ahead of a weak cold front that had been moving southeastward over Pennsylvania. Courtesy of NOAA/NESDIS.

boundaries (seacoast lines, mountain ridges, and the like such as seen in the satellite picture below). Often, they are triggered in lines ahead of cold fronts, as depicted in Figure 3.7 and in the satellite image in Figure 3.29. The latter shows a mass of clouds generated by the cumulonimbus clouds ahead of the cold front. Once established, the cumulonimbus can generate its own progeny. The inflow and outflow circulation into and out of the thunderstorms can act in unison and cause individual cumulonimbus cells to have a lengthy life span (sometimes as long as several hours or more), as the cool outflow from the clouds creates its own mechanism for lifting the warm moist air to saturation, forming newborn cumulus clouds.

In severe-storm situations both the direction and the speed of the wind change rapidly with height, as would be the case close to an upper jet core or jet streak. As such, the third factor referred to above is consistent with the presence of a strong horizontal temperature gradient near the surface, as discussed in Chapter 2. A clue that the atmosphere is organizing severe thunderstorms is that the latter tend to form themselves into rows, referred to as squall lines. Highly organized thunderstorms tend to form lines because the growth of new cells tends to occur along the right-hand flanks of the cumulonimbus cells facing the direction of motion of the cell. Such new cumuli development can continue irrespective of the trigger mechanisms mentioned above. Squall lines can reach lengths of hundreds of miles and persist for days.

Volumes have been written on the dynamics of thunderstorms and their characteristics, but the subject is beyond the scope of this book. What is important here is for the observer to recognize the development of thunderstorm clouds by noting the appearance of towering cumuli as their herald, as suggested by the towering cumulus clouds just east of the squall line in Figure 3.29.

Cumulonimbus clouds are the progenitors of hail, downbursts, tornadoes, and flash floods. They have a unique signature on weather radar. Unlike precipitation associated with sideways ascent, which appears as large swaths of grayish clouds on satellite imagery or as a uniform color (usually light or dark green coloring) on radar imagery, cumulonimbus clouds appear as small, bright, white pimples on color-coded satellite images or red spots on color-coded radar images, sometimes alone but often growing in squall lines or in clusters. On radar imagery, the reflectivity (strength of the return signal) is seen as a dollop of orange, red, or even purple showing where heavy rain or hail is falling, as in Figure 3.30. The steep sides of a cumulonimbus cloud are displayed as a very sharp contrast in colors on reflectivity. These features are especially noticeable when the clouds are penetrating the tropopause.

Squall lines and cold fronts

Squall lines, such as that depicted schematically in Figures 2.3 and 3.7, tend to form just ahead of cold fronts that extend from open-wave cyclones. Very severe and highly destructive thunderstorms frequently occur in the southeastern United States with this frontal type of pattern during the winter months, especially February and March. The reasons for this preferred location just ahead of the cold front are two-fold. First, the lift provided by the advancing cold front triggers the first convective clouds. The second reason can be understood best by looking at the wind directions in the warm sector, as shown in Figure 2.1. We notice that the air is flowing on the most direct path from the tropics into the warm sector just east of the front. This trajectory, which originates farthest south, brings with it the warmest air and, most importantly, the highest water vapor content. Squall lines themselves tend to move in the direction of the winds at middle and upper levels; for example, the upper winds would be from the northwest behind the three small squall lines shown in Figure 3.30.

FIGURE 3.30. A Weather Service Radar image of reflectivity values from Madison, WI, on May 28, 2012. Arrows denote a trio of small squall lines. Red areas denote the presence of cumulonimbus clouds. Courtesy of NOAA/ NESDIS.

Figure 3.31 shows the distribution of dewpoint temperature associated with the type of open-wave disturbance depicted in Figures 2.1–2.4. The salient of high humidity, the tongue of moist air with dewpoints between 15° and 20°C, extends northward just east of the cold front and coincides with the relative position of the squall line depicted in Figures 2.3 and 3.7. A general rule of thumb is that significant thunderstorms require dewpoints in excess of about 15°C and severe thunderstorms dewpoints in excess of about 22°C.

In general, the wind flow just ahead of the cold front moves almost parallel to the front on its left side. If we were to adjust the wind directions shown in Figure 2.1 to a coordinate system moving with the cold front, we would see that the air just ahead of the cold front is flowing closely parallel to the cold front, in this case from a southerly direction, as depicted in Figure 3.31. When the cold front extends into the Gulf of Mexico, as in the schematic example given in Figures 2.1 and 3.31, the air moving northward just east of the front possesses high dewpoints characteristic of the source region, the Gulf of Mexico. Alternately, a cold front trailing back into the Midwest or extending back into Canada would necessarily have much drier air ahead of it. By simply looking at the source region in this way we can assess the potential of the storm for producing heavy precipitation.

Stratocumulus: the hand-me-down cloud

When storms have spent themselves and fronts have swept through, the remnant moisture usually is seen as a stratiform version of the puffy cumu-

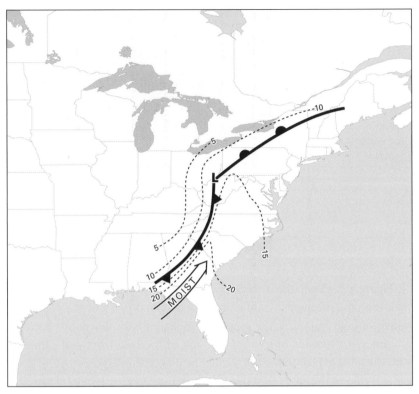

FIGURE 3.31. Distribution of dewpoint temperature in degrees C in the vicinity of the frontal system and warm sector for the schematic open wave cyclone discussed in Chapter 2, Figures 2.1–2.4. The core of the moist advection (the so-called "moist tongue") lies just ahead of the cold front and tends to move parallel to the front.

lus cloud. Stratocumulus clouds are nondescript and often cover the entire sky. The change in the thermal pattern in the lower atmosphere in the wake of a weather disturbance (cool air aloft and mild but moist air near the surface) promotes cumulus clouds to grow, while the large-scale sinking motions behind storms and fronts suppress the rising air currents and turn them into layers of flattened clouds. At first, they appear almost identical to their first cousins, the cumulus cloud, but after a short period of time these cumulus cloud lookalikes flatten out and typically turn into a gray layer of stubby cumuli.

Not infrequently, these clouds spread out to cover most or all of the sky, resembling their less benign cousin, the stratus cloud. In such instances the cloud could equally well be classified as stratus cloud, although the implications for future weather may differ between the two types of stratus. From

FIGURE 3.32. Images of stratocumulus clouds viewed from the ground.

FIGURE 3.33. Stratocumulus clouds as viewed from an airplane.

the ground, stratocumulus clouds seem gray and uninteresting (Figure 3.32). From above (Figure 3.33), these clouds appear as wads of mashed potatoes that have been spread thin across the landscape.

Nothing more than a sprinkle or flurry falls from these innocuous clouds, since they are shallow and seldom have much depth (or moisture content). Surprisingly, stratocumulus clouds may be the most common cloud type, since they cover thousands of square miles of the oceans where moistened surface air from mild ocean currents is made buoyant by the presence of cooler air aloft (Figure 3.34). In such cases, the ceaseless wind currents over the oceans also keep the lower atmosphere well mixed, promoting the formation of stratocumuli. Over land, stratocumulus clouds resulting from the spreading out of fair-weather cumulus clouds often disappear at sundown.

Stratocumulus clouds are often difficult to distinguish from their less benign harbingers of approaching inclement weather, the stratus clouds. If one were to wake up in the morning and see a sky such as that shown in Figure 1.9, it would be difficult at first glance to tell if this were a stratus cloud resulting from the progression of high, middle, and then low cloud in advance of a disturbance, or if it were an especially thick layer of stratocumuli that had spread from a more benign cumulus cloud origin, or the result of frictional convergence confined to the near-surface layers. One way to discern the

FIGURE 3.34. A visible satellite image showing the eastern Tropical Pacific covered by an extensive mass of cellular stratocumulus clouds. Courtesy of NOAA/ NESDIS.

FIGURE 3.35. Crepuscular rays emanating through a gap in a stratocumulus layer.

difference is to look for holes in the cloud deck. Stratocumulus clouds are often raggedy looking with some openings in the cloud revealing clear sky above, whereas true stratus, harbingers of stormy weather, tend to be more sheetlike and uniform and are accompanied by layer clouds at higher levels. If blue sky can be seen through these openings (as in Figure 3.32), the layer is probably a stratocumulus cloud, but if middle- and upper-layer clouds are visible, the low cloud is almost certain to be the kind of stratus cloud that heralds an approaching storm.

Viewed from a satellite image the latter is likely to be obscured by the middle and upper cloud, whereas true stratocumulus clouds are likely to be visible by itself as a deck of lower cloud, as is shown by the extensive area of cloud just west of Mexico in Figure 3.34.

Stratocumulus clouds may all look the same, but when sunlight bursts through a small opening, they can produce a brilliant display of crepuscular rays. What appear to be divergent rays of light breaking through an otherwise bland overcast is a good example of an optical illusion. The rays of sunshine are really parallel, and the dust and other large particles in the air help illuminate the sunbeams that have also been called Jacob's Ladder (Figure 3.35). Such stratocumulus clouds are usually the benign variety.

Clouds and the classic cyclone model

Let us now integrate the types of cloud discussed in this chapter with the classic cyclone model, exemplified by Figures 2.1–2.4 and by Figure 3.7, which shows the clouds associated with the classic comma-shape pattern, originally presented in Figure 2.3. Bear in mind, however, that the classic case is an idealization of a type of pattern that has an infinite number of variations.

With the aid of the classic case, however, we can imagine a multitude of scenarios as the disturbance approaches and later passes by the observer. Let us imagine observers in three different locations, initially located east of the cloud and weather system depicted in Figure 3.7. Scenario 1: the observer experiences the passage of the low pressure system just to the south. Scenario 2: the observer is situated well to the north of the low center. Scenario 3: the observer is situated to the south of the low center. These are now discussed in their respective order.

- **Scenario 1:** Initially, thin patchy cirrus (ci) are followed by more uniform cirrostratus (cs). The clouds will appear in the western and southern horizons but will seem to thicken most noticeably in the quadrant between west and south (see Chapter 6). Atmospheric pressure will fall slowly. Wind direction will vary between southwest and southeast. Clouds will lower to altostratus and altocumulus (as/ac) and then to stratus (st). Atmospheric pressure will fall more rapidly and precipitation will follow (ns), which are likely preceded by some virga and perhaps accompanied by fog if the observer is not far north of the warm front. Wind direction will remain in the southerly quadrant, between southwest and southeast, but may change to northeast with falling temperatures as the low moves by to the south of the observer. (As the storm matures, precipitation will spread farther to the west and southwest in the comma cloud.) Later, the precipitation will end and the cloud cover will be replaced by stratocumuli (sc), possibly accompanied by occasional light showers or snow flurries, and later by cumuli (cu). As the rain ends, the pressure will begin to rise and the wind direction will change from a southerly direction to a north or northwesterly quadrant.
- **Scenario 2:** Initially, thin patch cirrus (ci) are followed by more uniform cirrostratus (cs). The clouds will appear more in the southern and south-

eastern than western horizons but will appear to thickening most noticeably in the southern between south and southeast. Atmospheric pressure will fall slowly. Initially, clouds may continue to thicken to altostratus (as) and altocumulus (ac) and even to stratus (st) as in scenario 1, but winds will remain in a more easterly or southeasterly direction and the pressure will decrease only slowly. Precipitation may be light or nonexistent, and, subsequently, as the pressure begins to rise, the clouds give way to stratocumulus (sc) and cumulus (cu) as the winds change to a northerly and later northwesterly direction.

- **Scenario 3:** The process may begin with the appearance of thin cirrus (ci) and patchy cirrostratus, thickening from the west and northwest, with winds in the south to southwest. As the warm front moves over the observer, the sky will contain a mixture of patchy altocumulus (ac) and cirrus (ci), accompanied by cumulus (cu). Wind direction will change to southwesterly from southerly or southeasterly and the temperature will rise noticeably in the presence of more sun. The pressure may briefly rise but then later decrease slowly as the absolute humidity (the dewpoint) increases. As the cold front approaches, cumulus clouds (cu), accompanied by a mix of higher clouds, may appear to produce towers (tc) and the winds freshen from the south. In some instances a squall line may precede the cold front (cb), resulting in a thunderstorm, followed by some further squally and transient episodes of precipitation as the cold front passes. Winds are southerly but may come from the direction of individual thunderstorm cells when the latter are in the vicinity of the observer. Atmospheric pressure will rise abruptly after the passage of the cold front, and the winds will change to a more westerly and northwesterly direction, accompanied by a decrease in temperature and humidity.

An elaboration on these scenarios is presented in Chapter 6. The reader is invited to create his or her own scenario using the classic cyclone model. A typical variant, for example, is when a storm approaches the observer from the south, requiring one to somewhat rotate in the mind's eye the schematic weather maps shown in Figures 2.1–2.4 and Figure 3.7 in order to accommodate a storm track from south to north along the coastline. Warm fronts associated with coastal storms tend to parallel the coast, and thus the coastal storm track will also tend to move along the coastline toward the north. The procession of events described in the above scenarios will nevertheless be quite similar. As we will see in the next chapter, some storms can even approach from the northwest.

In the next chapter we will also look at some smaller-scale features of the weather in the northeastern United States that do not fall within the classic cyclone model. We will look at anomalous cases whereby storms come from a northwesterly direction rather than from the south or west, where cold fronts move from the east or north rather than from the west and where stratocumulus clouds alone can accompany heavy snowfalls.

CHAPTER 4
SMALLER-SCALE STORMS

A bit of review and a bit of history

The remarkable insight of the Norwegian scientists at the Bergen School of Meteorology just after the turn of the 20th century set the standard for a century of equally remarkable advances in the science of predicting the atmosphere. It took nearly 50 years to visually verify the Norwegian Cyclone Model when the Television and Infrared Observing Satellite (*TIROS-1*), was launched on April 1, 1960. Since then, despite countless iterations on this model published over almost the past 100 years, it is amazing how elegantly simple and true the Norwegian model still is today. The image in Figure 4.1 is from the Geostationary Operational Environmental Satellite (GOES-EAST/-13); it illustrates a classic mature stage of the Norwegian Cyclone Model and the comma-cloud pattern referred to in the previous two chapters (see its resemblance to Figure 2.8), with cold, warm, and occluded fronts superimposed on the image.

Essentially, most disturbances are a variation on the idealized Norwegian Cyclone Model. Figure 4.1 shows the classic and rather broad comma head extending into Canada but not unlike that shown in Figure 2.8. It is accompanied by cold, warm, and occluded fronts and possibly a secondary low forming at the intersection of the cold and warm fronts. The similarity of these systems to the classic model with its infinite variations underscores how important it is for our understanding to be familiar not only with the

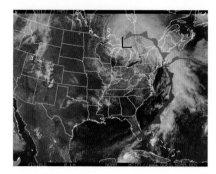

FIGURE 4.1. A classic Norwegian cyclone as seen from GOES-EAST visible imagery on May 2, 2012. Courtesy of NOAA/NESDIS.

idealized model but also with the local variations that occur as we focus on the eastern part of North America.

First principles of storms: a review

All midlatitude cyclones, except those associated with certain geographic features, such as desert heat lows, form along fronts. As we recall from the previous two chapters, a front is a boundary separating streams (formerly called air masses) of differing density, essentially temperature and density gradient. Fronts, however, are not an artifact of the human imagination but arise from fundamental dynamics governing the air motions. Recall that, in talking of density, we usually refer to cold air being denser than warm air as long as the comparison is made at the same pressure, although atmospheric density can vary slightly with water vapor content. Two main contributors to atmospheric density differences between airstreams are thus temperature and moisture. While it is intuitive that cold air is denser than warm air, it is not obvious that dry air is heavier than moist air at the same temperature, but this is very important in determining fronts, especially during the warmer half of the year. As a side note, the fact that moist air is lighter and less dense than dry air can be shown using the law of partial pressure discovered by John Dalton. The crux of the law is that mixed inert gas pressure (such as the atmosphere) is the sum of the molecular components of each gas (such as nitrogen, oxygen, or water vapor). Air with this mixture has a molecular weight of about 29 g/mole. Therefore, when more water vapor (H_2O with a molecular weight of 18 g/mole) is added to the atmosphere, the weight (pressure) decreases and the air becomes less dense. (Some fronts are just discontinuities in atmospheric density due to the difference in moisture content on either side of the front, but these types of fronts are relatively rare over the Northeast.) Henceforth, we will simply refer to temperature rather than density when speaking of frontal boundaries.

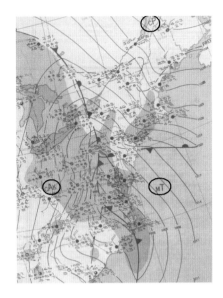

FIGURE 4.2. Surface weather map from 7:00 am EST, November 25, 1950. Air mass types as defined are circled. Courtesy the NOAA Central Library Data Imaging Project.

Although temperature is continuous across the front, the moisture and the temperature gradients are not continuous. The front's location can be ascertained by the temperature gradients on either side of the frontal boundary; indeed, this was once a formal method for determining the strength of a front. Fronts also have an impact on the pressure field, as can be observed on conventional weather maps by the presence of kinks in the isobars. These kinks represent the fact that fronts are also loci of lower pressure, with higher pressures on either side of it.

The largest contrast occurs between a very warm and humid air mass and a very cold and dry air mass. Over 50 years ago (e.g., Figure 4.2), when it was customary to label categorizations of air masses, this boundary would separate a continental polar air mass (cP) and a continental arctic (cAk) from a maritime tropical air mass (mT). At present, it is more appropriate to refer to these types of air masses as airstreams, in light of the fact that the air is constantly in motion, albeit originating in highly different regions that imprint the aforementioned labels on the airstream. Figure 4.2, which is presented mostly for historical interest, illustrates the way these air mass designations used to be plotted on weather maps. Maps like this were plotted by hand and contained an incredible amount of detail, including cloud types, weather, barometric pressure and tendency, temperature, dewpoint, wind direction, wind speed, and sky cover. Most of the stations located in the northeast corner of this figure were reporting cirrus overcast, a harbinger of impending stormy weather, as discussed in Chapter 3.

While the presence of a front is a necessary part of midlatitude cyclone development, it is not sufficient for cyclogenesis, as has been stressed in the previous chapters. This is where the three-dimensional aspect of atmospheric circulation comes into play. As we recall from earlier chapters, cold and warm fronts mark the boundaries of temperature gradients and of moisture itself, whereby these temperature (or density) gradients are manifested aloft, less by their discontinuous nature than by sloping pressure surfaces (the tilted surfboard) and the presence of a rapid increase with height of the wind speed; recall from Chapter 2 the discussion relating surface temperature gradients to the location of the jet stream. Recall also, that, despite the fact that horizontal temperature gradients tend to become weaker with increasing altitude up to the middle troposphere and beyond, the jet stream is often found not far from the surface location of the cold front (a little on its poleward side) and often oriented parallel to it. This linkage attests to the fact that the atmosphere functions as an organic whole, with all components acting almost simultaneously and synchronously.

However, as we have discussed, the winds within the jet stream are not uniform and contain numerous jet streaks with local wind speed maxima referred to as jet cores and accompanied by small wavelike disturbances rippling through this fast current. These weak upper shortwave disturbances are usually reflected at the surface by fast-moving but weak low pressure systems. Occasionally, a ripple (a weak low pressure system) will amplify (grow in size and intensity), and this will cause a shift in the wind field aloft leading to one of the many imbalances that occur in the atmosphere; this process is illustrated in Figure 2.12.

The shifting winds associated with low-level temperature advection lead to patterns of surface convergence and divergence. Since these surface effects are mirrored aloft by divergence (overlaying convergence) and convergence (overlying divergence), these diverging and converging airstreams at upper levels present the appearance of fast-moving air piling up in some sections and thinning out in adjacent regions—much like how heavy traffic that is moving quickly can slow down and lead to a mini traffic jam. The picture of the Champs-Élysées around the Arc de Triomphe in Figure 4.3 illustrates the sort of bottleneck that occurs when traffic is obliged to slow down where the road narrows, causing a convergence of cars. Once the road opens up, the traffic accelerates (and diverges), and everyone speeds away in a much lower volume of cars. (The analogy is limited, however. Unlike automobile traffic, air entering these upper convergent zones tends to speed up, although the flow is converging from the sides, thereby offsetting the effect of the speed

FIGURE 4.3. An early summer day in Paris looking northward along the Champs-Élysées.

increase. The opposite occurs in the divergent exit region of the bottleneck, where the air slows down but diverges laterally.)

These upper convergence and divergence patterns are merely reflecting events taking place throughout the atmosphere as the latter attempts to maintain a balance of forces. The atmospheric equivalent of a mini traffic jam leads to air at upper levels converging, or sinking, but these are reflected by divergence at the surface accompanied by rising surface pressure. Like the surface and its pressure tendency patterns, the flow at high levels also has its characteristic signatures of convergent and divergent patterns in the narrowing and opening of contour lines, whereby the opening up from congestion is the equivalent of divergence aloft, corresponding to convergence at the surface and surface pressure falls.

One may wonder how it is that surface convergence actually accompanies a fall in pressure if mass is being piled up in the convergent area. The answer to this seeming conundrum is that the divergence aloft is actually removing more mass from the column than is being introduced (converged) at the surface. The result is that surface air pressure still decreases, and rising motions occur with the attendant clouds and precipitation. The opposite occurs for surface divergence, whereby the upper-level convergence adds mass to the column despite its removal at the surface. This arrangement sometimes leads students of meteorology to incorrectly assign as a first cause processes occurring at jet stream level; in fact, as we have said, the entire atmosphere is almost simultaneously adjusting itself at all levels.

Figure 4.4 illustrates the three-dimensional nature of atmospheric processes occurring everywhere around the globe, including cyclogenesis, by depicting the aforementioned convergence/divergence dipole patterns (narrowing and widening of the flow) associated with an open-wave system.

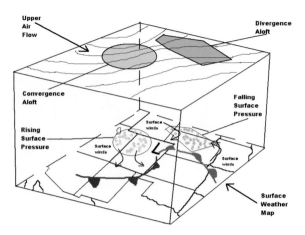

FIGURE 4.4. A three-dimensional view of the development of a Norwegian cyclone acting in consort with an upper-level disturbance. At upper levels, yellow denotes the convergence (reflected as divergence at the surface where there are surface pressure rises), whereas blue denotes high-level divergence (reflected as convergence near the surface, where surface pressures are falling).

The Norwegian atmospheric scientists, who figured out how the cloud pattern would look from above in the absence of satellites and without significant aircraft data, did an impressive piece of scientific detective work. Indeed, it was another 50 years before satellites were able to confirm their earlier cyclone model. However, we are now fortunate in that once the basic principle of storm formation is understood we can adapt this to any part of the midlatitudes and add a local twist to include more local phenomena. In the northeastern quarter of the United States, a classic example of this type of local pattern would be the Alberta clipper.

The Alberta clipper

One of the fascinating parts of studying the weather is that no two weather patterns (like snowflakes) are identical. No matter how similar the configuration of temperature, pressure, and clouds may be to a previous event, there are always important differences, and this is what makes predicting the weather such a wonderful challenge.

The Alberta clipper is a good example of a Norwegian cyclone that has been turned on its side. The same principles of rising and sinking motions are present, but since the movement of the developing cyclone is from northwest to southeast, the positions of warm and cold fronts are rotated about 90° (Figure 4.5) compared to the standard cyclone model.

FIGURE 4.5. An idealized depiction of the differing orientation of a typical cyclone compared with an Alberta clipper.

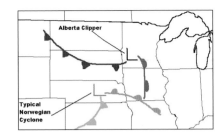

FIGURE 4.6. An infrared satellite view of an Alberta clipper crossing the Canadian Rockies on November 10, 2003. The top of the comma is circled. Courtesy of NOAA/NESDIS.

The term Alberta clipper is a mixed metaphor relating the origin location of this storm (the province of Alberta, Canada) with its speedy movement resembling a clipper ship. It usually moves southeastward from western Canada on a northwesterly flow to the west of an upper trough.

In its infancy, an Alberta clipper appears like a blob or comma-shape cloud moving across the Canadian Rockies (circled cloud mass in Figure 4.6). The rugged terrain and its distance from low-level moisture sources strip the disturbance to its core, which is reflected as wind speed maximum along the jet stream and a weak low pressure system at the surface with little precipitation; recall our argument from the previous chapter that relates the moisture content of a frontal system to the origin of the airstream at the tail end of the cold front, the part farthest from the low pressure center. In this case, the origin of the airstream is over southwestern Canada in Figure 4.5 and over the Rocky Mountains in the western part of the United States in Figure 4.6. Nevertheless, despite the relatively dry origins of the airstream ahead of the cold front, a sufficient amount of moisture exists to create the classic comma-shaped cloud system.

During the colder half of the year (which is Alberta clipper season), the prairies of Canada are often snow covered and the blanket of snow can also extend into the eastern plains. The temperature contrast between snow-covered ground and bare ground (and the warmed ocean surface to the west) tends to establish a temperature gradient and a northwesterly jet stream parallel to the low-level isotherms. This is similar to the presence of the so-called southern branch of the jet stream (mentioned in the previous chapter), which accompanies the strong temperature gradient between the Gulf of Mexico and the cold land area in winter, and acts as a steering guide for these storms.

Since the trajectory of these cold-season cyclones is mainly across the center of the continent, they tend to remain moisture starved. Three important sources of moisture exist for cyclones over the eastern half of North America: the Great Lakes, the Gulf of Mexico, and the Atlantic Ocean. Usually, the Great Lakes are too small, too cool, or even too ice covered to contribute much for the Alberta clipper's moisture supply and therefore to its precipitation area.

However, even with the Alberta clipper, a narrow ribbon of moist air from the Gulf of Mexico can still feed into the cyclone as it crosses the Midwest and upper Ohio River valley. This is suggested by the presence of the cold front starting to extend into the southern plains in Figure 4.7 (recall Figure 3.31 showing the moist tongue of air moving from the Gulf of Mexico ahead of a cold front). In this case, the air circulating counterclockwise ahead of the surface low and trailing cold front and west of the high pressure cell over the Atlantic Ocean is starting to direct moisture into the clipper from the lower Mississippi valley into its pathway, leading to more widespread precipitation. Figure 4.8, depicting this moisture flow on visible satellite imagery, shows that moisture is beginning to reach the clipper disturbance from the western Gulf of Mexico.

The interesting challenge for a weather forecaster in watching an Alberta clipper develop is to assess whether the storm is able to access a moisture source as it approaches the East Coast and to watch for possible secondary development along the coast in association with the coastal front and coastal temperature gradient discussed in Chapter 2. Not infrequently, the clipper can use the moisture and thermal contrast at the eastern seaboard to ignite the birth of a secondary disturbance along the coast into a full-blown nor'easter.

However, more often than not, there is a missing ingredient for an important snowstorm. This ingredient is usually a large cold dome of high pressure over New England that serves to bring cold air north of the approaching warm front, packing the isotherms together along the coast and creating a strong coastal temperature gradient. Unless it does trigger a coastal storm, the Alberta clipper itself typically deposits only a few inches or less of snow over the Northeast.

FIGURE 4.7. The surface weather map from 7:00 am EST, November 11, 2003, depicting an Alberta clipper (circled area and arrow), which has moved southeastward over the Great Lakes from its position shown in Figure 4.6. Green areas represent precipitation, the dashed blue line the 0°C isotherm. Courtesy of NCDC Map Archives.

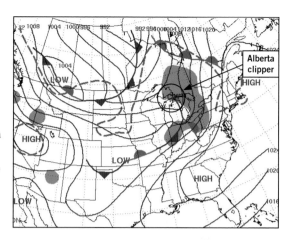

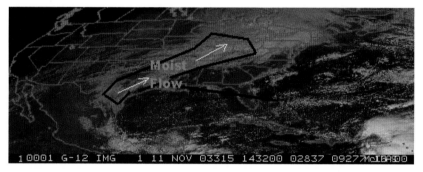

FIGURE 4.8. A visible image from GOES-EAST on November 11, 2003. Courtesy of NOAA/NESDIS.

The exciting part of watching an Alberta clipper is to anticipate its triggering of a coastal storm. The great blizzard of '78 possessed the common missing element, referred to above, so that it developed into a historic New England nor'easter. One can see from the weather maps on February 4 (Figure 4.9) that a high pressure system resided northeast of the weak frontal system. This high pressure system corresponded to a dome of cold air. Initially, the clipper was in South Dakota (Figure 4.9), and, a few days later, it had slipped under the Northeast cold dome (the high pressure cell), generating a secondary low pressure center along the coast. The latter rapidly intensified into a nor'easter (Figure 4.10).

The original low pressure center of the clipper, shown in Figure 4.9, persists as a weak troughlike feature west of the new low center shown in Figure 4.10. (The absence of precipitation over the ocean in Figure 4.10 is merely due to the absence of observations. In reality, the precipitation shield would have

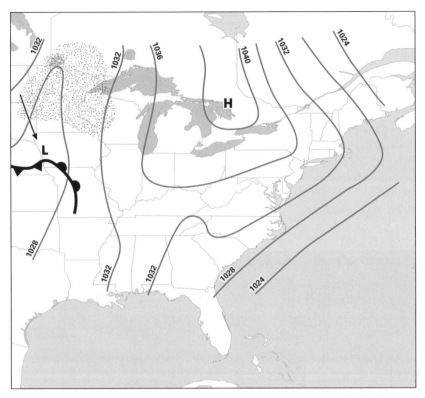

FIGURE 4.9. The location of an Alberta clipper, February 4, 1978 (arrow), which led to a developing nor'easter later on February 7 (Figure 4.10). Shading denotes sustained precipitation. Contours are isobars of surface pressure (mb).

extended farther east and north of the warm front.) As discussed in Chapter 2, the storm's subsequent path closely paralleled the coastline.

Several books and numerous publications have been written on this well-heralded snowstorm. For a quick summary along with a slide show, visit the National Weather Service website (see the Appendix).

The classic nor'easter

The term *nor'easter* refers to the direction from which the wind is blowing that is experienced at coastal locations when this type of storm occurs along the Atlantic seaboard. Mariners noted the wind as blowing from any of 36 points on a compass and used the term *nor'* for "north." While nor'easters can occur anywhere in the country, they are favored in the east and their ferocity is manifested most frequently along the eastern shoreline of the continent where the northeast flow north of the storm center circulates unhindered

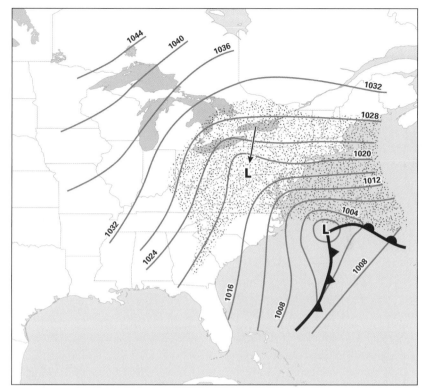

FIGURE 4.10. A developing nor'easter on February 7 that began with an Alberta clipper (Figure 4.9). Shading denotes sustained precipitation. Contours are isobars of surface pressure (mb). The location of the original Alberta clipper is marked with an arrow.

across the adjacent ocean and piles up waves and tides along the coast. This onshore wind also advects moisture inland, thereby adding to the precipitation rates in these coastal regions.

It was Benjamin Franklin, in Philadelphia, who noted from correspondence with his brother in Boston that although the surface winds blow from the northeast, the clouds, rain, and snow actually migrate from the southwest. He had noted that clouds and rain in Pennsylvania obscured a lunar eclipse in November 1743, while his brother in New England enjoyed viewing much of the eclipse before clouds and rain ensued. The winds first blew from the northeast at Philadelphia and then stronger in Boston later on. By his reasoning, it seemed that storms had a three-dimensional structure, which was quite an insight into these complex storm circulations.

The intense interest in nor'easters focuses mainly on their wintertime character that can produce enough snow and ice to bring the Atlantic Sea-

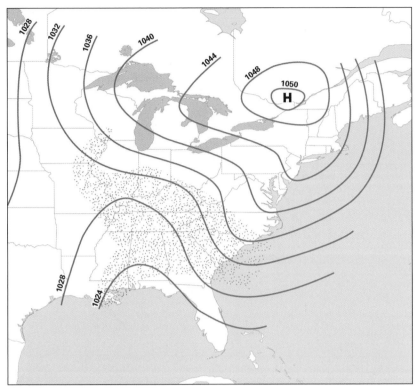

FIGURE 4.11. A surface weather map depicting a large cold air mass over the Northeast at the inception of the President's Day storm on February 18, 1979 that is just forming along the Gulf of Mexico coastline (see also Figure 2.11, which is relevant to this storm). Shading denotes precipitation (mostly in the form of light snow, except rain over the Gulf of Mexico states). Central pressure of the high pressure system was 1050 mb (approximately 31.00 in. of mercury).

board activities and traffic to a crawl. When it comes to snowstorms, there are three key components. Reasons for the cyclogenesis are described in the previous chapters. Here, we refer to the nor'easter itself and its abundant snowfall; if any of the following three precursors are missing, a classic snowstorm or nor'easter is unlikely to occur. The following summarizes our previous discussions pertaining to these types of coastal storms.

1. Cold air. This may seem obvious, but it is an essential part of any snowstorm. There are nor'easters that have produced a cold rain in January across the Mid-Atlantic region and ones that have brought snow in October and April! More often than not, there is a missing ingredient in an otherwise ideal situation for the formation of a nor'easter; this is usually a large cold dome of high pressure over New England. The depth and position of the cold air are very important. Ideally, a fresh batch of polar air with its associated cP (continental polar) high pressure system must be in place or expected to arrive as the cyclone develops. The high pressure represents a dome of cold, dry air for mild, moist air to be lifted in a slantwise fashion (as initially described in Chapters 1 and 2) from an ocean source. This lifting is directly related to warm-air advection between the warm sea air and the cold-air dome. The larger the high pressure system to the north of the cyclone (i.e., the higher the central barometric pressure), the larger the precipitation area. In fact, there is a crusty old rule among weather forecasters that says, "Predict the high, and then predict the storm." The example in Figure 4.11 is from February 18, 1979, the day of the first President's Day blizzard (see Figure 2.11 for the track of this storm and its relationship to the coastal front). This storm is typical of many intense coastal cyclones in that they tend to form along the coast of the Gulf of Mexico.

2. Moisture. Unlike cold air, where surface-based temperature and pressure are easily quantifiable descriptors of the cold dome, we need to rely on a remote-sensing tool like satellite imagery to determine the richness of the moisture supply. One method is to use a satellite product. The satellite can measure the total moisture in the atmosphere and display this in a channel, called a water vapor channel. When combined with another channel, or window of energy being sensed in the infrared wavelength, the source region of a plume of deep moisture can be readily displayed. The image in Figure 4.12 illustrates the supplemental use of the infrared satellite imagery by showing that the superstorm of March 1993 had its moisture feed originating from the tropics of Central America. (This source would be at the lower end of the cold front, which coincides with the tail of the comma cloud and

March 13, 1993

Deep
Moisture
Source

FIGURE 4.12. A full disk infrared satellite view from a GOES satellite. Courtesy of NOAA/NESDIS.

the southern end of the cold front.) The presence of the cold dome would have aided this moist flow by contributing a flow of moist air from the ocean surface to the east directly into the comma head north of the warm front.

3. A trigger. While a full explanation lies beyond the scope of this book, it can be shown mathematically from an analysis of Newton's equations of motion for a rotating fluid that the rate at which cyclonic rotation increases depends upon the magnitude of the existing cyclonic rotation: the larger the latter, the larger the former. This constitutes the theoretical basis for the exponential increase of cyclonic rotation with time, manifested as the slantwise instability associated with explosive cyclogenesis. It follows that a favored location for initial cyclogenesis is an existing cyclonic vortex, such as a weak low pressure system or a front, preferably in the southern branch of the jet stream. By contrast, however, the rate of increase of anticyclonic rotation in the presence of divergence decreases with increasing magnitude of the anticyclonic rotation, thereby making anticyclogenesis an impossibility, as discussed in Chapter 2.

Many times in the cold season, a series of ill-defined wiggles in the two branches of the jet stream current will be predicted to magically come together and produce cyclogenesis, a nor'easter. The disturbance may be marked at upper levels only as a well-marked ripple (wave) in the jet stream that corresponds to a couplet of rising motions ahead of it and sinking motions behind (see Figure 4.4), such as might be occurring in conjunction with a weak frontal system and surface low. This would be represented by the Alberta clipper described earlier in this chapter. These ripples can serve as seeds for cyclone growth and, as precursors to explosive cyclogenesis, are warily followed by forecasters.

While the Alberta clipper or similar disturbances in the northern (mid-continental) branch of the jet stream may ultimately be able to tap into the rich supply of moisture that is adjacent to older fronts farther south, as indicated in Figure 4.8, a strong jet stream over the continent (the so-called northern branch) can displace the significant cold dome, which we have indicated is a necessary condition for a snowy nor'easter, while keeping aforementioned precursors south of the northeastern states. Therefore, a key ingredient to a nor'easter is often the presence of the southern branch of the jet stream and a weaker northern branch.

That the disturbance often originates in association with a vigorous southern branch of the jet stream is illustrated in Figure 4.13, where a separate and intense wave in the upper-air flow is shown coincident with an incipient surface low forming along the Gulf Coast or near the Florida Panhandle.

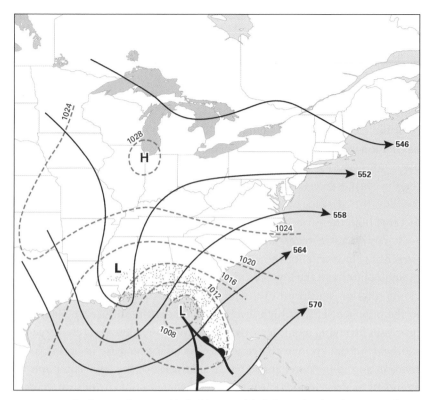

FIGURE 4.13. Surface weather map (dashed lines are labeled as isobars) and upper-air flow (the 500-mb pressure surface in decameters above sea level) at 7:00 am EST, January 4, 1987. The 500-mb height contours are approximately equivalent to the isobars at the level of about 550 dm (5500 m) above sea level. Higher heights on the 500-mb surface correspond to higher pressures on the 550-dm surface. Shading denotes sustained precipitation. Barbs at the end of the height contours denote the direction of the geostrophic flow at that level. The intensifying surface low is located in the strong southwesterly flow ahead of an upper-level trough that had been displaced southward over the Gulf of Mexico by the southern branch of the jet stream, reflected in the narrow spacing of the height contours in that region.

Conversely, the northern branch of the jet stream was relatively weak on this occasion. The figure illustrates an important point first raised in Chapter 2, which is that the upper trough (the solid lines in Figure 4.13) must lag to the west of the surface disturbance (the dashed lines in Figure 4.13) in order for warm and cold advections to take place at low levels around the storm, and therefore for cyclone development to take place. The similarity of the storm's location in Figure 4.13 to that in the earlier stage of the cyclone development shown in Figure 2.10 underscores that while no two storms are ever alike certain patterns tend to repeat themselves in very similar ways.

This storm eventually moved up the Atlantic coast and produced a swath of 10–20 in. of snow along and just east of the Appalachians.

There are favored pathways for precursors to nor'easters. Figure 4.14 shows the paths of some precursor storms as well as paths that failed to produce a nor'easter. The reader is invited to decide which color of arrows would be best for a nor'easter.

If the colors red, green, and blue (RGB) were selected, then the reader has made a correct choice and now understands how nor'easters get started! Note, however, that these three trajectories all turn northward when reaching the coast, in consonant with the discussion of why coastal storms tend to follow the coastline when they experience explosive development and the fact that both the surface and upper-air patterns are locked together.

It may strike the reader upon inspection of Figure 4.4 that low-level convergence responsible for the increase in cyclonic rotation northeast of the cyclone center and the low-level divergence causing the increase in anticyclonic rotation southwest of the low center must correspond to rotations in the opposite sense at high levels from that at the surface. Indeed, dual processes involving upper-level divergence overlying low-level convergence northeast of the cyclone center and its attendant increase in anticyclonic rotation and upper-level convergence southwest of the low center and its attendant increase in cyclonic rotation in that area would cause an enhancement of both the ridge north of the surface warm front and the trough west of the surface cold front, as suggested in Figure 2.12. In cases of explosive cyclogenesis the ridge and trough can become so highly distended toward the north and south, respectively, that they deform the westerly current so much that the upstream progression of troughs and ridges is halted. Like a highway accident that blocks traffic, the eastward movement of troughs

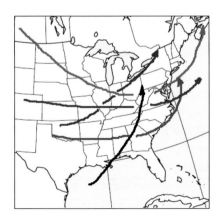

FIGURE 4.14. Paths of low-pressure systems, some of which produced nor'easters.

and ridges is often blocked by explosive cyclogenesis. The movement of the wave train (and therefore the weather) is affected for some considerable distance both upstream (to the west) and downstream (to the east) of the area of cyclogenesis. We will make further mention of blocking patterns in the next chapter.

Backdoor cold fronts

Most houses have a front door and a back door. Visitors are usually received at the front door. In a similar way, the atmosphere sends visitors our way that change our weather. We sometimes call these weather visitors storms or fronts. The majority of fronts arrive through the front door (or from the west or northwest). But on some occasions, the change in the weather can come in the back door (from the north or the east). There are certain locations for backdoor fronts that occur because of the regional topography, the presence of the Eastern Seaboard of the United States, and the presence of cold ocean temperatures east of the Canadian Maritime provinces.

As we recall, a front is a boundary between airstreams of differing density and/or temperature gradient. The Atlantic Ocean provides a natural air mass boundary, not only during the winter but also during the spring and summer months as well, when water temperatures (and subsequently the temperatures of the air above the water) are considerably colder than the heated land to the west—a reversal of the wintertime pattern. During the sultry days of the summer, this cool oceanic air provides a local cooling effect at the beaches when the breezes blow onshore. Figure 4.15, evocative of the chimney or bonfire effect mentioned in Chapter 1, illustrates the sea breeze circulation, which is being driven by the pressure difference between a cool ocean and a heated land surface—higher pressure at the surface over the ocean.

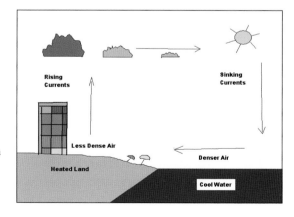

FIGURE 4.15. An illustration of the sea breeze circulation that can locally mimic a back-door cold front.

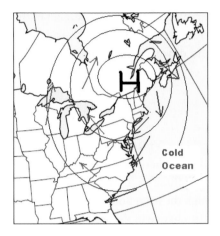

FIGURE 4.16. The typical circulation around high pressure in the Northeast that leads to a back door type of cold front east of the Appalachians.

The sea breeze is the basis of a backdoor cold front, but it occurs on a much larger scale. During the period from March through July, when high pressure moves or grows across the Northeast or the Canadian Maritimes, its clockwise wind circulation will channel ocean-chilled air inland, giving the effect of a cold front coming from the east. The surface weather map in Figure 4.16 schematically illustrates this relatively common configuration.

Occasionally, a cold front that is oriented east–west across southern Ontario and Quebec will accompany a mound of very cold air developing over the snow-covered regions of northeastern Canada or the cold waters of the Labrador Current (Figure 4.16), and cold air from this high pressure system will move southward along the east coast of Canada as a shallow cold front. The circulation around the burgeoning high will propel the front southward and westward into New England and the Mid-Atlantic region with remarkable surface temperature contrasts. Should the wind be blowing strongly from the west preceding the passage of the front, then warming due to air subsiding from the Appalachians will enhance the thermal contrast between the warm air ahead of the front and the cool sea air. Note that the position of the high in Figure 4.16, so favorable for a backdoor cold front, is virtually identical to the optimum position for the nor'easter, described previously (Figure 4.11).

A remarkable case of this type of backdoor cold front is shown in the surface map from April 16, 2003 (Figure 4.17). Differences in temperature and weather across the backdoor front can be quite profound, as is illustrated in this figure.

The four stations circled in Figure 4.17 draw our attention to the contrasting temperatures and wind directions across the front. South of the front the temperatures are above 27°C (80°F), whereas they are 5°–6°C (41°–43°F) or

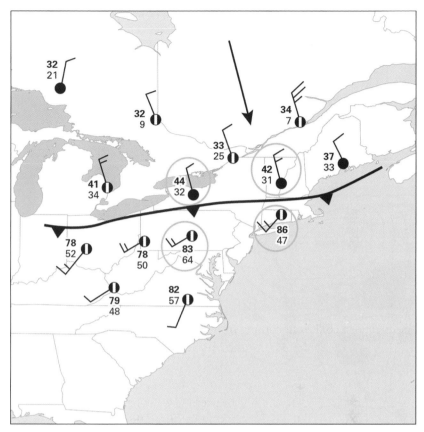

FIGURE 4.17. A regional surface weather map from 2:00 pm EDT, April 16, 2003. Cloud cover is represented by the degree of black filling inside the station circles. Temperatures in degrees Farenheit (F) are plotted above and just to the left of the station circle. Wind barbs are represented as in Figure 2.1. The heavy arrow shows the direction of the northerly wind blowing from an area of nearly freezing temperatures toward the front, south of which the winds are from the southwest. Dewpoint temperature in degrees F is plotted just below the temperature. The two pairs of circles situated north and south of the cold front draw attention to the warm–cold contrast across the front, whereby the temperatures decrease by about 22°–25°C (40°–45°F) from south to north.

lower north of the front. These two pairs of temperatures located on either side of the front illustrate an unusually strong contrast between the warm air on the south side of the front and the cold air on the north side of the front, separated by a relatively small horizontal distance—a 22°–25°C (40°–45°F) change in less than 100 miles (160 km). Stations north of the front are mostly cloudy, and a few of these stations farther north are reporting light precipitation (not shown). Such contrasts in temperature and weather, and

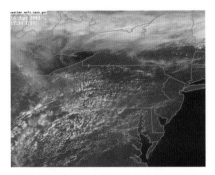

FIGURE 4.18. A visible satellite imgery from GOES-EAST at 1:31 pm EDT, April 16, 2003. Courtesy of NOAA/NESDIS.

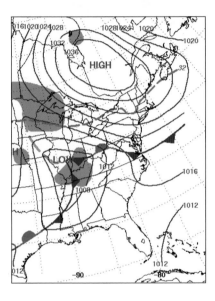

FIGURE 4.19. April 17, 8:00 am EDT, approximately 18 hours after the time of Figure 4.17, showing the same backdoor cold front, which has pushed well south of Pennsylvania. Courtesy of NCDC Map Archives.

their attendant implication for human comfort, underscore the importance of being able to predict the occurence and location of backdoor cold fronts.

The visible satellite image (Figure 4.18), zoomed into the Mid-Atlantic region at about the same time as the surface analysis in the previous figure, shows that most of the cloud cover and some precipitation trails tens of miles behind (on the cold side of) the front, which coincides approximately with the northern border of Pennsylvania. This is typical of a backdoor cold front since the chilly air is quite shallow near the front, so that the lifting of warmer, moistened surface air blowing from the south (overrunning the cooler air to the north) is strongest some distance inside the cold air, thereby allowing condensation and clouds to form well north from the surface position of the front in the area of stratiform cloud. Note the contrast in Figure 4.18 between trails of cumulus clouds in the warm air and the bank of stratiform clouds north of the front.

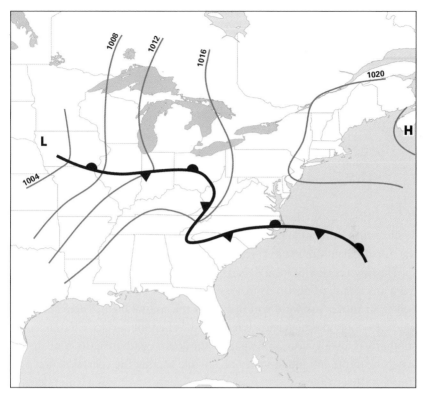

FIGURE 4.20. A surface weather map illustrating a late spring backdoor cold front on June 11, 1998.

Backdoor cold fronts frequently descend from the north over Pennsylvania and New York State or move from the east over New England, although the latter type of backdoor front occurs only a few times each year when they move inland and westward as far as Pennsylvania. In this particular case, rather strong northerly winds (shown in Figure 4.17) were able to move the front southward into Virginia where it finally stalled as the next disturbance approached (Figure 4.19).

The season for backdoor cold fronts along the Eastern Seaboard extends from March into July when the temperature contrast between the ocean and the land and (earlier) between snow-covered Quebec and the warmer land surface to the south is greatest. As summer approaches, this type of front can be a true spoiler for those vacationing at the beach. Several days of low clouds, drizzle, brisk onshore wind, and readings staying in a narrow range, usually in the 50s or 60s (10°–15°C), sends some beach revellers home early. Despite the cloud cover, backdoor fronts and fronts generally arriving from the north or east are seldom accompanied by significant rainfall. The weather map in Figure 4.20

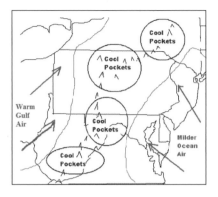

FIGURE 4.21. A depiction of cold air wedged into the mountain valley region of the Mid-Atlantic states.

shows a classic example of a salient of ocean-cooled air having reached as far south as the Carolinas in mid-June where its westward progress is impeded by the Appalachian Mountains (Figure 4.21)

The way a backdoor front departs is typically by the surface winds turning from an easterly or northerly to a more southerly or westerly direction. The cooler air moves eastward and then exits the coastal plain. There are times when the chilly air's departure is less rapid. When the surface winds become southerly (blowing from some part of the south quadrant), the cool air can get stuck in the mountain valley area while leaving the coastal region and the Ohio River valley.

A strenghtening breeze from the southeast on the east side of the Appalachians will draw milder, cooler air from the Atlantic, and a fresh wind from the southwest will bring in notably warmer air from the Ohio River valley, as is the case for Figure 4.21. Where these two air currents converge over the Appalachians, there is rising motion, considerable cloud cover, and even some evaporational cooling due to precipitation. As a result, the cool air shrinks into pockets until it only occupies valleys within the complex terrain of the central or northern Appalachians, as is illustrated in Figure 4.21.

Reluctant warm fronts

Let's return now to a discussion initiated in Chapter 3 regarding how warm fronts sometimes appear reluctant to cross the Appalachian Mountains in wintertime, when the retreating cold air becomes trapped in the valleys. Often, the surface warm front in such situations is drawn south of its real location at an altitude of 300–400 m above the valley floor. Figure 4.21 illustrates how retreating cool air in advance of a disturbance becomes trapped in the many valleys of the northern Appalachians. Figure 4.22 shows a case where the warm front during December 2012 remained just south of Pennsylvania, while continuing

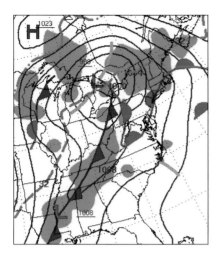

FIGURE 4.22. Warm front drawn south of Pennsylvania due to the entrapment of cold air at the surface within the many valleys in the Appalachian Mountains. The suggested position of the warm front above mountaintop level may be coinciding with the southern border of the (green) precipitation area located just north of Pennsylvania. The two blue lines are surface isotherms (−18°C and 0°C [0°F and 32°F]). Courtesy of NCDC Map Archives.

to move northward over western Pennsylvania and New England. In this case the actual location of the warm front at 300–400 m above the surface may be the southern border of the precipitation area in this figure.

Often, as in Figure 4.22, the warm front will advance across the western part of Pennsylvania, Virginia, West Virginia, New York State, and eastern Ohio, forming a narrow salient of much warmer air between the cold front and the cold intrusion pushing southward to the east. While the surface front remains back to the south in the mountainous areas, the front at mountain-top level continues to move northward.

Lake-effect snow

For winter weather enthusiasts, lake-effect snowfall can bring a blizzard to their backyard while having little impact on areas just a few miles away. The regions just downwind of the Great Lakes are the most prolific snow produc-ers east of the Rockies and often have abundant snow cover during the many years that the remainder of the Northeast is snow starved. The amazing at-tribute of lake-effect snow is that it usually occurs in a so-called fair-weather pattern, typically with northwest winds. Whereas nor'easters are large cy-clones that envelop many states at the same time, a lake snow event often happens with mainly sunny, blustery conditions east of the Appalachians.

There are four basic ingredients for lake-effect snow:

- Cold enough air in contact with warmer and, of course, unfrozen open lake waters
- Air temperatures below freezing

- Winds that blow across the lake, preferably along the major axis of the lake
- An upper disturbance that enhances the rising motions, although this is not necessary

The first ingredient has two components. The most important is the difference between the lake water temperature and the cooler air at approximately 5,000 ft (1.5 km; approximately at 850 mb) above it. The threshold difference is found to be about 13°C (24°F) in order for the lower part of the atmosphere to undergo rapid overturning due to cold air overlying warmer water and, hence, warmer air in contact with the surface. Therefore, if Lake Erie's water temperature is 5°C (41°F), in order for lake effect to occur the air temperature at 5,000 ft would need to be −8°C (17°F) or lower. Differences greater than 13°C or 24°F will encourage deeper overturning and larger snow squalls.

The secondary component in producing lake-effect snow is that the mean temperature of this 5,000 ft (1.5 km) air column should be below freezing; otherwise, rain showers instead of snow squalls would occur.

There are unusual occurrences of lake-effect snow as early as mid-October, such as in Buffalo, NY, on October 12–13, 2006, when trees were still in full leaf. The lake temperature was still 12°–13°C (54°–56°F), but the air temperature aloft was less than −10°C (14°F), resulting in a heavy, wet snowfall. The amount of snow for so early in the season was quite amazing (Figure 4.23).

Generally, lake-effect snow season starts around Halloween and ends by April Fool's Day but there is a period of time, highly variable from year to year, between late January and the end of February, when the lakes are largely frozen. After these dates, the air is not cold enough aloft to cause the enhanced evaporation from the surface, whereas little moisture can be evaporated from the water when the lakes are covered with ice. By far the heaviest snow squalls occur before the lake begins to freeze.

For Pennsylvania, only Lake Erie lies in the pathway of frigid air masses capable of producing exceptionally heavy lake-effect snows close to Lake Erie. Being the shallowest of the Great Lakes, Lake Erie freezes over about 90% of the winter seasons.

The crossover points (Figure 4.24) in late July and early March mark the beginning and end of the lake-effect season. The process of lake effect can best be seen in Figure 4.25 as bubbles of warm (orange arrows) and moist (green arrows) air being drawn upward by mixing with cold, dry air. During its path across the lake, the mixing process deepens and the bubbles of more

FIGURE 4.23. Lake-effect snow totals (in inches for October 12–13, 2006). Courtesy of the National Weather Service office in Buffalo, NY.

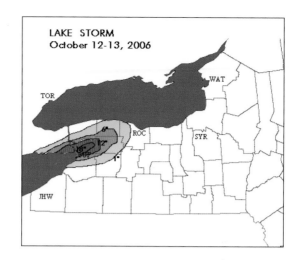

LAKE STORM
October 12-13, 2006

FIGURE 4.24. A generalized view of lake versus air temperature. The word stable refers to the season when the lake temperature (blue trace) is below the air temperature (red trace), thereby inhibiting evaporation. Precipitation occurs when the lake temperature is above the air temperature; this situation will produce rain when the temperature is above freezing over the lowest 1.5 km and snow when the temperature over the lowest 1.5 km is below freezing. Courtesy the NOAA Central Library Data Imaging Project.

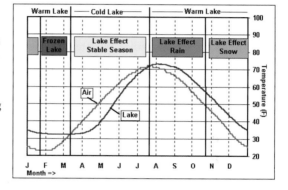

FIGURE 4.25. A generalized view of lake effect as described in the text.

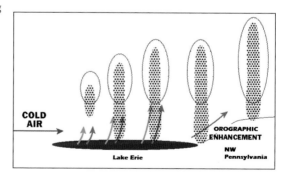

buoyant air expand and grow into towering cumulus clouds (see Chapter 3 for a description of these clouds). The combination of surface friction and higher terrain on the lee side of the lake forces the air to ascend even farther, resulting in heavy snowfall. Clearly, this process is completely a result of cumulus

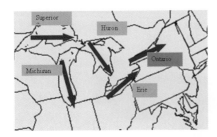

FIGURE 4.26. Optimum wind directions for lake-effect snowfall and the associated lake names.

convection and not the large-scale stratiform precipitation associated with frontal boundaries and slantwise ascent as described in previous chapters.

The trajectory of the cold airstream across the Great Lakes is crucial in determining the location and amount of snowfall. Certain trajectories favor more snowfall. In the map shown in Figure 4.26 regions prone to the heaviest snow are shown at the end of the longest pathway across the lakes.

One could argue whether a reverse direction (easterlies) would produce the same effect on the other side of the lake. The answer is that air flowing from the east or southerly directions tends to be warmer than air flowing from the west or northwest, at least in winter. The result is that air appearing over the lake surface with easterly or southerly wind directions would be milder than air moving from the west or north, thereby reducing the lake surface–air temperature difference and giving rise to a weak lake effect or none at all.

It is worth noting that the pair of trajectories of winds for Lakes Michigan and Huron (corresponding to the southward directed arrows in Figure 4.26) resemble each other, as do the pair of trajectories for Lakes Erie and Ontario (corresponding to trajectories directed toward the northeast), meaning that significant lake-effect snow often occurs in pairs on these lakes. Optimum wind directions occur when the wind direction is parallel to the long axes of the lakes. Significant snowfall close to the lake shores can nevertheless occur when the winds are blowing perpendicularly to the lake axis, notably over northern Wisconsin, where a cross-lake flow still must traverse considerable distance over Lake Superior. Even in the case of Lake Erie, the smallest and shallowest of the lakes, cross-lake winds can still deposit significant snowfalls just downwind from the lake shore.

Two other terrain factors serve to focus the snow squalls: terrain elevation and surface friction. The former is obvious. When the air residing over the lake surface reaches the shoreline, it is forced to ascend tens to hundreds of meters in elevation over the higher land surface. Heaviest snow is therefore most likely over the relatively narrow zone between the lake and the higher ground downwind.

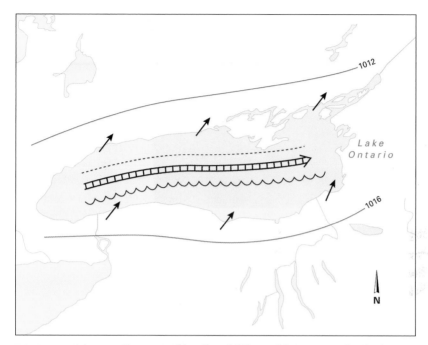

FIGURE 4.27. Schematic illustration of the effect of differential friction on surface-level convergence for air blowing parallel to the long axis of a lake, such as Lake Ontario. The direction of the surface-level wind over the lake (hatched arrow) is almost parallel to the isobars, but it is flanked by surface wind components on either side over the land, each having a greater angle across isobars toward lower pressure than the surface wind over the lake. The result of this arrangement is to have a zone of frictional convergence along the south side of the lake (squiggly line) and a zone of divergence along the north side of the lake (dotted line).

The second factor, friction, has been mentioned previously, notably in regard to Figure 2.14 and in this chapter to Figure 4.25. In the latter figure, the frictional effect occurs when the air reaches land and is obliged to slow down, causing a convergence in the airstream and induced ascent along the shoreline.

Friction also affects precipitation patterns around the lakes in another way. As Figure 2.14 demonstrates, a shallow layer of air near the surface, being affected by surface friction, does not behave in geostrophic balance. Rather, the addition of friction force to the mix of pressure gradient and Coriolis forces requires that the wind direction exhibit a component of motion across the isobars toward lower pressure. Over a relatively smooth water surface the frictional drag force is much smaller (typically about one-third to one-half of that over a rougher ground surface).

Figure 4.27 illustrates the effect of this difference in frictional drag on air over water and over land. Air flowing along the lake axis over the water,

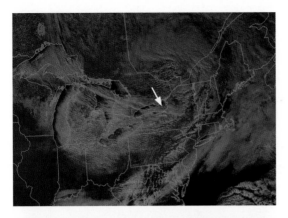

FIGURE 4.28. GOES visible satellite imagery (top) showing the snow bands trailing downwind from the lakes. A radar image for the same time (bottom) shows precipitation bands over and downwind of Lakes Erie and Ontario. The white arrow on the bottom denotes the major bands marking air trajectories crossing several of the lakes; that on the top points out the location of Lake Ontario. Courtesy of NOAA/NESDIS.

being subject to less friction, moves almost parallel to the isobars (higher pressure on the right side of the motion), while air moving over the land surface exhibits a significant component across isobars toward lower pressure. The result is a line of convergence along the right side of the lake (with respect to the wind direction), causing upward motion and precipitation to favor an alignment near the south shore of the lake, while a corresponding line of divergence and sinking motion tends to inhibit precipitation along the north side of the lake.

From a satellite and radar perspective (Figure 4.28), lake-effect snow events are easy to distinguish. The visible satellite imagery from GOES-EAST shows dozens of parallel bands of cumulus clouds marking the air trajectories, some with pathways across several Great Lakes. The latter are more noticeable because they are wider and deeper (left frame). The single radar image from the National Weather Service office in Buffalo (bottom frame) illustrates the so-called string bean nature of one lake-effect band aimed at this city. Although the flow is generally not along the long axes of Lake Ontario (white arrow on the top panel of Figure 4.28), the largest area of radar echoes occurs just downstream from this lake because the trajectory of the winds creating these bands traversed several lakes. The small white arrow on the bottom frame denotes the terminus of these lengthy snow bands that swept across Lakes Superior, Michigan, and Huron before arriving over Lake Ontario. While the amounts of snow immediately downwind from the lakes are not infrequently in excess of a foot or two, bands of lighter precipitation extend well inland, producing light amounts (typically less than an inch) over the central interiors of Pennsylvania and New York State.

Nocturnal cooling and daytime heating

We return now to a discussion of the temperature and static stability changes occurring near the ground as the result of radiational cooling at night and solar heating during the day. These changes, illustrated in Figure 4.29, will help to illuminate the discussions of buoyant convection first introduced in Chapter 2 (see also Figure 2.7b) and further developed in regard to cumulus convection in Chapter 3. It will also help in understanding the next couple of related topics presented in this chapter.

Starting with the near-sunset sounding (the thin solid line in Figure 4.29), the initial lapse rate, as the result of the previous day's heating by the sun, is somewhat less than 10°C per kilometer decrease in temperature with height. Recall that the decrease in temperature with height that an unsaturated parcel would exhibit upon being lifted is 10°C per kilometer. As the evening proceeds, cooling due to long wave radiation emission from the surface to space begins to form a shallow temperature inversion (temperature increasing with height) close to the ground, illustrated by the dashed line. This inversion deepens with time, so that by early morning it typically extends up to a few hundred meters (the thin pecked line); above the inversion, the original near-sunset sounding remains unchanged above that shallow layer.

Nocturnal cooling occurs because longwave (thermal) radiation from the ground to space causes the temperature at the ground and, increasingly, in the air layers just above the ground to cool. Because the atmosphere is largely transparent to thermal radiation, the greatest cooling occurs at the surface, above which the air is cooled indirectly by a very limited turbulent mixing. (Indeed, the inversion acts as a strong brake on vertical exchanges of air parcels.) The inversion is actually most pronounced at the ground level and weakens with height. It is not unusual at sunrise for the temperature at grass-top level to be several degrees below that observed when measuring with a thermometer placed at eye level outside one's back door. Frost may even form during the night on a lawn when the air temperature via thermometer reads 3°–4°C (36°–38°F).

Nocturnal cooling will occur only to the extent that the thermal radiation emitted from the ground is allowed to escape to space. Cloud cover can impede the radiational losses to space from the ground; a complete cloud cover will prevent an inversion from forming. Cooling near the ground (and therefore the strength of the inversion) is also reduced on windy nights when turbulent eddies created by the vertical wind shear serve to mix the cooler air below with warmer air above, even in the presence of an inversion. In general, the calmer the winds and the clearer the sky, the deeper and more pronounced the inversion. The inversion will also tend to be weaker on hilltops where the radiationally cooled air can slip down into lower elevations. It is not unusual in hilly regions for an observer located at the base of the hill to experience a cool breeze blowing down from the side of the hill just after sunset; this phenomenon is referred to as cold-air drainage. Valleys may therefore become cooler than hilltops during clear nights.

Shortly after dawn, the effect of solar radiation reaching the surface begins to warm the air, quickly erasing the nighttime inversion. By early afternoon the heat added to the atmosphere has raised the temperature over a layer adjacent to the surface, thereby creating a new vertical temperature profile (the heavy solid line in Figure 4.29). In the absence of condensation, turbulent mixing of the air creates a lapse rate of about a 10°C per kilometer decrease with height, the same rate of cooling experienced by a parcel of dry air being lifted without condensation. Within this surface mixing layer (here, shown to be about 0.7 km in depth), the lapse rate is 10°C per kilometer, except over a shallow layer near the ground where the lapse rate is greater than 10°C per kilometer. This very steep lapse rate close to the ground is the result of solar heating being strongest at the ground and the fact that the proximity of the surface impedes efficient mixing of the air. Just as radiational cooling is maximized at the surface during the night, solar heating during the day has its greatest effect on the lapse rate adjacent to the ground. It is not uncommon on a sunny day for the temperature to be several degrees higher at grass-top level than that measured by a thermometer outside the back door.

It is easy to see from the figure that an air parcel, upon being lifted during the day at the cooling rate of 10°C per kilometer (dotted line with arrow), remains warmer than the surroundings and therefore positively buoyant (with an upward buoyant force) for some distance above the surface, and it is thus able to form a cumulus cloud where the vapor in the thermal reaches saturation; cloud base is shown to be at a level of about 0.8 km in the figure.

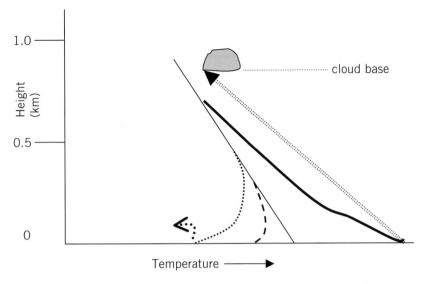

FIGURE 4.29. Schematic temperature versus height profiles, showing the vertical distribution of temperature at several times during the day: near sunset (thin solid line); near midnight (dashed line), near dawn (thin pecked line), and later during the late morning (heavy solid line). The dotted line with arrow denotes the path of a positively buoyant thermal rising from near the surface and cooling at a rate of 10°C per km of ascent, remaining warmer than the existing temperature sounding upon reaching condensation at the level labeled (cumulus) cloud base. The short dotted line with arrow to the left of the dawn sounding represents a negatively buoyant pocket of air, which is unable to rise from the surface through the nocturnal inversion layer because it remains colder than the environment upon being lifted. Note that it is understood that all the soundings join the near-sunset sounding, which remains temporarily unchanged above the levels where the nocturnal cooling or daytime heating and mixing occur.

Depending on the profile of temperature above cloud base, the cumulus cloud can continue to grow vertically. Clearly, daytime heating promotes buoyant convection within and above the mixing layer. Similar to the nocturnal case for radiational cooling impeded by cloud cover, cumulus are less favored when the sky is overcast because surface heating is reduced.

By contrast, however, a parcel of air lifted vertically at night (the very short curved dotted line on the left, designated by a small arrow) will be much cooler than the surrounding air, and so sinks back rapidly. Consequently, as just mentioned, the presence of the nocturnal inversion greatly inhibits convection and can only be overcome by the presence of strong winds. Not surprisingly, therefore, a field of fair-weather cumulus or stratocumulus is likely to disappear at sundown. With the presence of the nocturnal inversion and the inhibition of turbulent mixing (in windless conditions), frictional drag tends to reduce the wind speed, which tends toward zero near the surface on calm, clear nights. Conversely, the wind picks up during the morning hours when faster-moving air aloft is mixed downward. This is often quite noticeable later during the morning or early afternoon, when turbulent mixing of the air due to solar heating of the ground brings down faster-moving air from above. The layer over which this mixing occurs is that where the lapse rate above the ground is equal to or greater than 10°C per kilometer.

One interesting phenomenon, only briefly mentioned here, is that the presence of a nocturnal inversion, and the quick removal of daytime mixing at dusk, can lead to the formation of a low-level wind speed maximum (called a low-level jet) just above the nocturnal inversion layer, sometimes attaining speeds of up to tens of meters per second. This effect is most pronounced over flat terrain where such a low-level wind speed maximum is unimpeded by topography. The low-level nocturnal jet is thought to be important for the formation of severe thunderstorms over the Midwest.

Mountain valley effects

The lay of the land—the local terrain—plays a significant role in altering the overall weather. While mountains and valleys themselves do not necessarily create or destroy storms, the way the air is diverted around and through these topographic features does change the intensity of local weather, especially thunderstorms. Gaining an understanding of the typical wind flow in and around hilly terrain will assist in the interpretation of how storms morph as they reach one's backyard. An example of the mountain effect resembles the sea breeze circulation referred to earlier in this chapter.

The inland version of a sea breeze circulation

Each day, particularly during the warmer half of the year and for a few hours after sunrise or later in the day nearing sunset, when skies are clear, the sides of the hills facing the sun are warmed more than the valley floor because the sun striking the hillsides makes a less oblique angle than on a horizontal surface. This leads to the differentially heated ground, which when warmed also warms the air above it and establishes a rising air current. The air from the valley floor moves upslope to replace the rising current in the form of heated, buoyant bubbles that convey evaporated moisture from the surface. Eventually, a small cloud may form directly over the ridge top, and a short-lived countercurrent of air descends into the valley. Such is the mountain valley circulation by day (Figure 4.30). The steeper the terrain, the more pro-

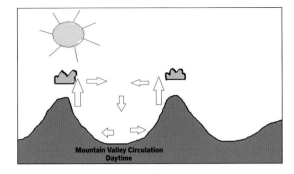

FIGURE 4.30. Schematic illustration of the mountain valley circulation during the day.

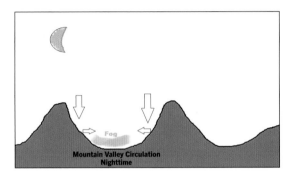

FIGURE 4.31. Schematic illustration of the mountain–valley circulation at night.

nounced the effect. These types of circulations provide convenient thermal chimneys upon which many birds, such as hawks, love to soar.

Along the front range of the Rockies from May through August, this process is so repetitive that one can almost set a watch by the time that the first cumulus appear over the mountaintops. In the East, cumulus rows tend to line up along the Appalachian ridges in mid- to late morning. Subsequently, the stronger winds aloft disrupt the local circulation and push the mountain clouds in the same direction as the winds, which is often from the west. However, when the air mass is rather warm and moist and the winds aloft are weak, as is typical in July and August, the ridge top cumulus can blossom into cumulus congestus (towering cumulus) and even into cumulonimbus before being nudged along by higher winds aloft. Not surprisingly, thunderstorms often first form along a natural boundary such as a ridgeline. This effect is especially noticeable over the White and Green Mountains of New Hampshire and Vermont in summertime.

At night, a reverse circulation occurs (Figure 4.31). Under clear skies with nearly calm winds, the hilltop and hillsides will cool more quickly than the valley due in part to the wooded surfaces at the higher elevations and because the cool air tends to drain away down the sides of the hills and accumulate in the valley floors. Its appearance in the valley is palpable as a fresh and cool evening breeze to those who live at the base of these ridges. In some spots, where the valleys are small and bowl shaped, the accumulated cold air can even reach near-freezing temperatures during the early morning on clear days in summertime.

As the night progresses, and especially during the lengthening darkness of autumn, the collected cool air reaches the dewpoint and initially forms dew and, in time, fog. The fog is first seen near sources of moisture, such as creeks, streams, or even water filtration plants where the humidity is locally higher than in the surroundings. On some occasions, the cool, moist air does not form fog until just after sunrise, which is the time of minimum temperature for the day. The slight warming that occurs somewhat later after sunrise causes a mixing of the lower atmosphere wherein the moist air near the surface is brought up to saturation level in a rising thermal, resulting in a foglike cloud that quickly dissipates. This process is essentially the same as that which allows us to see our breath on a cold morning.

Rain as affected by mountain ridges

Unlike the song from the classic movie *My Fair Lady* that claims the rain in Spain falls mainly on the plain, in reality, most of the rain and snow falls on the windward (upwind) side of hills and mountains—even in Spain. Terrain influences, therefore, can significantly affect the distribution of precipitation. The effect is most pronounced during the summer when migrating thunderstorms receive an extra boost crossing a ridge, as the air currents associated with the convective storm are forced upward with greater vigor as they approach and begin to ascend a ridgeline (Figure 4.32). However, much to the

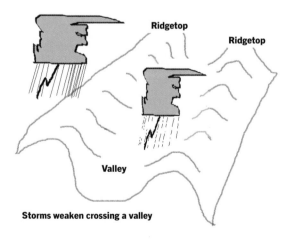

FIGURE 4.32. Schematic illustration showing how uneven terrain influences thunderstorm precipitation. In this figure, the wind flow is from left to right, and the storms form over the windward sides of the ridges and tend to dissipate on the downwind sides of the ridges.

repeated disappointment of storm watchers and gardeners in the downwind valleys, the same thunderstorm often weakens as it descends the ridge and traverses the lower terrain, causing reduced rainfall in the valley. This effect is especially noticeable in the valleys of central Pennsylvania, just downwind from the Allegheny Plateau.

In larger-scale storm circulation, the air flow is also disrupted by the topography. As the winds shift when the storm moves by a mountain/valley region, the rain or snow may completely cease, allowing sunshine to break through gaps in the clouds caused by localized sinking air motions.

An observer can be guided by the clouds

Ultimately, the sky itself will signal to the observer the nature of the mountain valley circulation. A morning that begins with fog foretells that cooler air has drained downhill from nearby mountains and is being collected in the valley to a sufficient depth and chill to cause widespread condensation and stratus or fog to be formed, sometimes in the form of a roll cloud, as shown in Figure 4.33. A day that sees puffy cumulus clouds appearing over the ridge top by noon informs an onlooker that the heated hillsides dominate the wind circulation. Even in the midst of a large winter storm, one can be witness to the modifying effects of local terrain, when the clouds suddenly

FIGURE 4.33. Downslope wind causing clouds to form into a long roll cloud in the valleys of central Pennsylvania.

part as the approaching disturbance causes a wind shift resulting in a local downslope wind flow from a nearby mountain ridge.

In the next chapter we will look at the way professional forecasters use some of the latest techniques and computerized model results to make their forecasts. Finally, in the last chapter we will integrate much of what has been presented into a discussion about the practical aspects of weather forecasting, with the hope that readers, using the tools of observation and comprehension, can serve as their own weather forecasters to make reasonably accurate predictions of the weather up to a day or so in advance.

CHAPTER 5

THE FOUNDATIONS OF WEATHER FORECASTING

Good observations: a philosophical view of forecasting

The best way to begin the forecasting process is by getting in touch with the weather. The more aware a person is of the weather, the wind direction, barometric pressure, and the sky, the better he or she will be able to utilize a weather forecast and even to make predictions. As such, getting in touch with the weather means more than just looking at the sky, although that is both a very good starting point and a necessary element in understanding and predicting the weather.

Although this book is aimed primarily at the amateur, we'll discuss going further than making observations and predictions at home for those whose interests lie beyond the basics or perhaps for the beginning meteorology student. To expand beyond the professional or amateur forecaster's line of sight requires access to weather data streams. It could be as straightforward as a current radar or satellite image of an area or even a country, such as those presented in previous chapters. These products help one to augment the visual observation and thus to gain perspective using the myriad data sources available.

Important sources of weather information are computerized weather forecasts, usually expressed in terms of weather maps 6, 12, or 24 hours apart, projected into the future 5–10 days ahead. Beyond a day or two, these predictions are much better than any human can make unaided by computer models, although the skill in these predictions does deteriorate with time from the present and the forecasts are valid only in the sense that they are able to approximate the locations of the highs and lows, troughs and ridges beyond 5–7 days from the present. Yet, human awareness of the sky and weather and intelligent use of the model forecasts are essential for making an improvement over the model results.

It was once thought that improved mathematical models, improved data coverage, data accuracy, and data access could ultimately lead to so-called perfect forecasts. We now realize that this goal is not only impractical but mathematically unattainable, as will be discussed later in this chapter. At present, no model is able to predict with any skill beyond about 12 days. Before computers, reliable forecasts could not be counted on beyond 24–48 hours.

In the short run (0–48 hours), however, human skill is still essential for improving predictions from weather forecast models. The most basic type of short-range forecast is something called persistence, which is simply to predict a continuation of the current conditions. A similar type of forecast is called climatology, which assumes that the weather will be that of the normal weather for that day, as derived from past weather data. In order to represent improved real skill of the forecast, however, beyond persistence or climatology, weather forecasts must be modified by the observational and integrative skills of the forecaster. Besides applying the principles discussed earlier in this book, being acquainted with the peculiarities of the local weather and microclimate and being informed by local temperature, wind, and pressure measurements will greatly inform the forecaster as to how the weather may be changing over the next few hours or more. Such predictions are known as nowcasting, which is essential in forecasting very short-term weather events, such as blizzards or thunderstorms. Indeed, the best short-term forecasts for a given region are very frequently made by local forecasters.

There are many other ways to get in touch with the weather, such as to have an overview of the various scales involved in changing the weather. Consider the following model: Len Snellman, one of the founders of the National Weather Association (NWA) for professional forecasters proposed in the 1970s a so-named forecast funnel as a way to gain an excellent perspective on the forecast process. The top layer, planetary scale, refers to weather systems that span a good fraction of the latitude circle, tens of thousands of kilometers. Synoptic-scale disturbances, the highs, lows, troughs, and ridges we see on weather maps, span scales of thousands of kilometers. Mesoscale features are typically tens to hundreds of kilometers in size; local-scale varia-

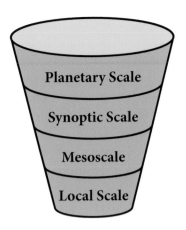

FIGURE 5.1. Forecast funnel. Courtesy of the National Weather Association.

tions in weather are on the scale of a few kilometers or less, such as differences between a mountain and valley weather, between urban and rural parts of a city, and even between neighborhoods. Synoptic-scale features are discussed in Chapters 2 and 4. Mesoscale and local-scale features are discussed in Chapters 3 and 4. We now address features on the planetary scale.

Making use of upper-air weather maps

Much like retrieving a tennis ball in the surf, in order to ascertain where the ball is going, one needs to understand the medium in which it is embedded. There are wavelets that cause it to bob up and down, and there are larger waves that can cause the ball to lunge forward or be carried farther from shore. But these waves are also driven by the phase of the local tide, whether it is coming in or going out: in other words, waves within waves within waves. The likeness to the Snellman funnel can be seen as the tennis ball representing one's location, which is embedded in an environment with many small perturbations (wavelets—analogous to local scale), which is also dominated by larger waves (analogous to synoptic scale) and by the tide (analogous to planetary scale and seasonal changes). To gain perspective in the forecast process, it is necessary to begin with the bigger picture. This can be both temporal and spatial. Snellman was primarily focusing on the spatial perspective. Here, we look at a hemispheric view of current weather systems, the planetary scale, and steadily zoom down to the local scale.

The planetary scale covers most of the Northern Hemisphere in Figure 5.2, which is a near-hemispheric representation of the 500-mb pressure level—equivalent to a map with isobars at a constant elevation, about 5.5 km above the surface. Planetary waves that consist of the ensemble of alternating ridges and troughs are embedded in a basic westerly flow. (Recall from

Chapter 2 that the wind direction will lie parallel to the contour lines, higher pressure on the right, with stronger winds corresponding to more closely spaced contours and that troughs and ridges at this level reflect the presence of cyclonic or anticyclonic vortices superimposed on a westerly current; troughs are therefore equivalent to low pressure systems and ridges are equivalent to high pressure systems.)

Synoptic-scale waves generally vary in length between 2,000 and 10,000 km from trough to trough. The smaller end of this spectrum is referred to as short waves. Short waves are interesting because they travel faster than longer waves and therefore accompany rapid changes in pressure and temperature at the surface that are conducive to cyclogenesis. Longer waves move more slowly and sometimes become stationary or even retrogress backward toward the west. However, like fast-playing golfers, short waves can usually pass through larger and more slowly moving waves. Short waves can also move slowly when they are either of large amplitude or appear to be cut off from the westerly flow, essentially resembling a nearly isolated spinning vortex. Such isolated vortices, cut off entirely from the westerly flow, may sit in place for days. Formation of an isolated vortex at these levels may occur as the result of cyclogenesis, which may cause the trough to become highly distorted, as discussed in Chapter 2. In the absence of a definite westerly current, a lone vortex at these levels has little or no steering current in which it can move. A clue that the vortex is beginning to move is that it starts to become asymmetric, much as a spinning top may start to move laterally, albeit erratically, if one were to stick a large piece of gum on one side. Cyclogenesis,

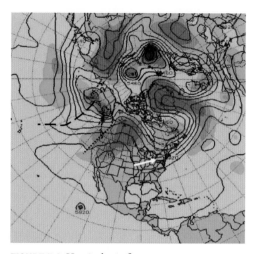

FIGURE 5.2. Hemispheric flow pattern as represented by the heights of the 500-mb surface (contour lines approximate the isobars on a map at about 5.5 km elevation) and its anomalies (reddish colors indicate above average and bluish colors below average for the day of the year). The white arrow denotes the location of Pittsburgh, Pennsylvania, referred to in the text. The heavy dashed line denotes the axis of a long wave trough in the Pacific, and the thin pecked line to its east marks a short wave trough embedded within the longer wave. Courtesy Penn State's e-Wall.

as noted in the previous chapter, can also lead to the amplification of the downstream ridge and the next downstream trough, causing a ripple effect that can extend for thousands of kilometers. Current computer models can usually deal with these kinds of processes, but it is nevertheless important for forecasters to watch out for them.

Sometimes troughs in these longer, more slowly moving or stationary waves appear to be made up of a cluster of shorter wave troughs, which give the appearance of one very long wave. The latter may appear to shift eastward or westward in response to the passage of shorter wave troughs moving through the larger one.

Let's look at one of the troughs on the 500-mb pressure surface shown in Figure 5.2, starting with the one situated over the northeastern part of the United States, whose axis lies near the white arrow. Anomalously cold temperatures associated with below-normal heights of pressure surfaces are located in this trough, primarily poleward of the jet stream—the bundle of isobars that are relatively close spaced. At the location of the arrow, the temperatures are close to normal. Conversely, above-normal temperatures are prevalent in the ridge, notably equatorward of the jet stream.

Short waves usually accompany surface disturbances, the latter being located on the east side of the trough axes, as depicted in Chapter 2 and in Figure 4.13. Short waves are of great interest to meteorologists because they are favored for cyclogenesis. Shortwave disturbances typically move eastward at a speed of several hundred kilometers per day, roughly with the average speed of the westerly current at 500 mb. For this reason, the 500-mb level is sometimes referred to as the steering level, because the surface disturbances and the 500-mb wave seem to be moving eastward with the average westerly wind at that level, although the winds are not actually steering anything. (Small-scale features such as the Alberta clipper, referred to in Chapter 4, and even hurricanes do tend to move with a direction and speed very close to that of the 500-mb winds in which these features are embedded.)

Now let's look at the Atlantic Ocean east of the Canadian provinces. There, we see a very large amplitude (and warm) ridge extending into the polar region, which will move very slowly, if at all; for that reason, this type of pattern is called an omega block, so named because of its shape resembling the Greek letter omega and because its slow speed and large amplitude tend to block the eastward movement of waves upstream—those situated to its west. Such systems tend to have a weak, poorly defined westerly current and are not especially favorable for cyclogenesis, despite their prominence on the weather map.

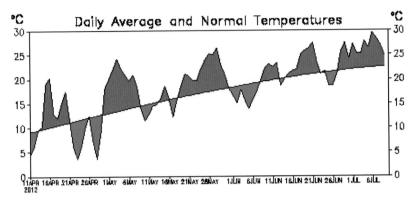

PITTSBURGH, PENNSYLVANIA

FIGURE 5.3. A 90-day temperature anomaly trace for Pittsburgh. Red areas represent above-normal temperatures and blue below-normal temperatures. Courtesy of NOAA's Climate Prediction Center.

Conversely, small-scale waves with relatively short wavelengths, typically 1,000–2,000 km from trough to trough (or ridge to ridge) and possessing smaller amplitudes, move more rapidly, typically with the speed and direction of the winds at 500 mb. Within the longer waves, one can often identify these smaller waves—wiggles that tend to move faster than the parent short wave. The heavy dashed line just west of the continent in Figure 5.2 denotes the trough axis of the longer, synoptic-scale wave, whereas the thin pecked line marks the axis of a very weak trough embedded within the longer wave system. These shorter wavelength and smaller amplitude systems tend to be reflected at the surface by weak disturbances, such as the Alberta clippers. They move very rapidly in the direction of the upper winds, while the parent wave moves much more slowly toward the east, allowing the smaller wiggle to outrun the parent wave, move through the ridge of the larger wave, and then move into the next downstream, synoptic-scale trough. An example of such a feature is again that associated with the Alberta clipper. These wiggles are still classified as synoptic scale, albeit on the lower end of the size spectrum of atmospheric disturbances.

One way to appreciate the passage of these waves and wiggles within the concept of the Snellman funnel would be to plot a time series (in temperature, pressure, etc.) for one's location and over the past few months up to the last hour's reports. Figure 5.3 shows how such a representation helps identify periods of above- and below-normal temperatures and therefore helps with the recognition of current anomalous situations. It also illustrates the cyclic passage of both large-scale and small-scale troughs and ridges in the upper flow.

Thus, by following the progress of the large and small synoptic-scale waves, which tend to maintain their identity and have lifetimes of weeks, one can easily project whether the location will experience below- or above-normal temperatures. For example, consider the location of Pittsburgh at the head of the arrow in Figure 5.2. Pittsburgh is located near the axis of a moderately long wave trough wherein the jet stream is displaced toward the south close to that city; farther north, the temperatures at the surface are below normal (the blue area). Since these longer synoptic-scale waves move more or less at a constant speed toward the east (and sometimes very slowly), the anomalously cold or warm periods may persist for several days or longer, resulting in the station experiencing a possibly lengthy cooler or warmer period, the duration of which could be ascertained by inspection of the predicted location of that trough on computer forecasts. In the case of Figure 5.2, we would expect Pittsburgh to be experiencing near-normal temperatures, if not slightly below, judging from its location in the trough, although it is somewhat south of the jet stream flow. Passages of long and short waves are evident by the various wiggles in the fluctuating time series of Figure 5.3.

Adding value to current forecasts

There are two adages among veteran weather forecasters that have withstood the test of time: (1) the trend of the model over a period of time provides an improvement to forecasting over that based on the current model prediction, and (2) the consensus (or average) of a number of forecasts of the same event or meteorological parameter (e.g., temperature) is superior to any individual forecast of that event or parameter. The second statement is borne out by many forecasting contests among meteorology students. If 10 meteorologists were to make 10 forecasts of, say, the maximum temperature for the following day, the average (consensus) of those 10 forecasts would always do better than most individual forecasts on any given day, and the average of that average over a week or a month would almost always be superior to the average of any individual forecaster over that time period. Even for the nonspecialist, listening or reading (in the newspaper or on the Web) to as many different forecasts as one can obtain, by mentally averaging the results, will result in a better overall forecast than any individual prediction.

The superiority of consensus gives rise to forecasting using an ensemble approach. A forecast ensemble contains either the same prediction model from the same initial time but with many slightly different starting (initial) atmospheric conditions or several different prediction models forecasting for the same area and time period. Consensus is another way of employing the old adage that two heads are better than one, or in the counsel of many is wisdom in achieving a superior method of weather prediction. One of the most spectacular examples of the superiority of consensus forecasts is that of the infamous and highly destructive Hurricane Sandy, which struck the New Jersey coast at the end of October 2012. Consensus forecasts employing the average hurricane track from several different models were able to successfully predict the correct track and point of landfall by this storm as much as a week in advance.

There are many ways to display an ensemble forecast. One type of ensemble model, used to adjust the forecast, is sometimes called the model of the day, which uses trends and involves tuning the computer-generated forecast maps according to a perception of their accuracy each day, as judged by their consistency in time. For instance, one might wish to predict the weather for the coming weekend for the northeastern United States, a week or more in the future. It would be reassuring to have confidence in such an extended forecast. In particular, one would like to know the time of arrival or location of a front or area of precipitation, initially indicated for that weekend in the computer-generated forecast map five or more days ahead. Such a feature might not even present on the weather map at the initial time of the forecast, although the forecast map may show it to be over the targeted area days later. An educated guess as to the reliability of the forecast, however, might be provided as follows by a series examination of the computer-generated forecast maps, one day at a time.

For example, suppose that early on Monday morning, July 23 (Figure 5.4), a storm watcher is concerned about the weather over Pennsylvania during the coming weekend of July 27–29. Computer-simulated maps made at 0000 GMT (2000 hours eastern daylight savings time the previous day) on the 23rd indicate the passage of a low pressure system and front over the northeastern United States on the 1200 GMT (0800 local time) forecast map for Friday morning, the 27th, 4.5 days into the future (labeled accordingly in Figure 5.5). Although a prediction this far in advance may be reasonable, there is clearly some uncertainty in a prediction 108 hours into the future.

How much can we trust this forecast? To increase confidence in the forecast (or at least to allow an assessment of its credibility), one can apply the model-of-the-day technique to this example. In this case, the model-of-the-day method is employed by looking at trends in the successive predicted

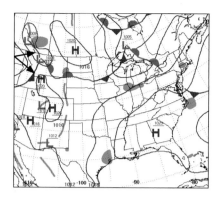

FIGURE 5.4. Actual surface weather map for 8:00 am EDT, July 23, 12 hours after the initial forecast time. The boldfaced arrow in the northwest corner denotes the low pressure center and trailing cold front whose location is to be predicted four days ahead. Courtesy of NCDC Map Archives.

locations of the weather pattern—the cold front and low pressure center—over the northeast at 1200 GMT (0800 local time) on the 27th, starting with the 4.5-day predicted map published at 0000 GMT on the 23rd; Figure 5.4 shows the actual map 12 hours later. At that time the front and low pressure system in question are still located over the western states. Each day, as a new set of extended forecasts is published, we repeat this assessment of the forecast for the weekend, but closely monitor the changes in the predicted features for the target area and time as the time gets closer to the weekend.

For the purpose of demonstration, we will illustrate the model of the day by each day plotting the essential features of the weather map over the Northeast, as seen on successive forecasts for the target region, the Northeast. We do this at successive 24-hour intervals, first showing the locations of these features predicted at 0000 GMT on the 23rd using the 4.5-day forecast; on the 24th using the 3.5-day forecast map for 1200 GMT on the 27th; on the 25th at 0000 GMT for the 2.5-day forecast for 1200 GMT on the 27th; and on the 26th at 0000 GMT for the 1.5-day forecast for 1200 GMT on the 27th. In so doing, we look closely at the succession of maps for our target time and region, specifically at the weather pattern on each successive forecast map. We look for how consistent the model is in establishing the locations of the relevant features and what biases the model is displaying in that weekend forecast.

Let's see how this procedure works in practice. Consider Figure 5.4, which shows the actual surface weather map and sea level isobars at 1200 GMT on July 23, 12 hours after the initial forecast time of 0000 (2000 local time the previous day; not shown) on the 23rd. Let's just look at the low pressure and frontal systems (arrows on the map), then located over the western United States and predicted to arrive over the Eastern Seaboard 4.5 days later (4 days after the time of Figure 5.4). We plot the weather pattern over our target region, the northeastern United States, specifically the predicted locations

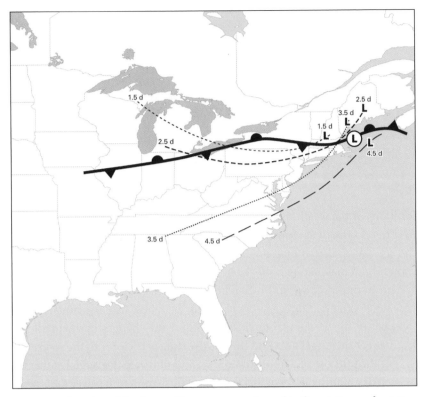

FIGURE 5.5. Location of the observed low pressure center and trailing stationary front at 8:00 am EDT on July 27, 2012 (denoted in the conventional manner). Predicted locations of this low center and trailing fronts predicted for this date and time on successive days, 4.5 days, 3.5 days, 2.5 days, and 1.5 days in advance, are shown and labeled accordingly.

of this low pressure system and its attendant front. We do this for successive days—the 4.5-, 3.5-, 2.5-, and 1.5-day predicted maps and the locations of these features at the same predicted time, 1200 GMT on Friday, the 27th (Figure 5.5).

Accordingly, the 4.5-day forecast shows the cold front and low pressure system, previously located over the western United States, reaching the Atlantic coast at 1200 GMT on the 27th. Each successive forecast for the same time (1200 GMT, July 27) places the low center in about the same location, along the Atlantic coast. We conclude that the prediction of the surface low appears to be quite consistent, and therefore as the weekend target time gets closer, we can have confidence in its predicted location.

The cold front trailing south of the low, however, appears to be displaced increasingly farther north with each successive forecast, approximately coinciding with the observed location of the front from 2.5 days ahead until the weekend target date. We would have less confidence in its predicted location.

What the forecaster can learn from this type of analysis is that the low center location, being highly stable in its successive positions in the forecast, is likely to have some credibility in its predicted location along the New England coast. In moving the front northward with each successive forecast, the trend suggests that the model is trying to adjust the location of the front northward until the 2.5-day forecast. However, the front's predicted location seems to have stabilized in the same location as the 2.5- and 1.5-day forecast, suggesting that the model had initially put the front too far south but finally made up its mind to place the front close to where it was later observed 2.5 and 1.5 days in advance, which is just north of Pennsylvania. The forecaster has confidence in the placement of the low center and could conclude as early as 1.5 days ahead of the 27th that the front would be stabilized north of Pennsylvania. In practice, however, the forecaster might wish to perform this type of analysis at 12-hour intervals and even further ahead than 4.5 days. Similar model-of-the-day forecasts can be made for precipitation.

Ensemble forecasts: the wisdom of consensus

In 1967 an MIT meteorologist by the name of Edward Lorenz published a paper that revolutionized the science of weather prediction. He discovered, in trying to run a very simple climate model on his computer, that if the initial conditions for the model were changed ever so slightly, the simulated weather patterns would quickly diverge from the previous simulation. Lorenz's paper completely changed our thinking about predictability, which was subject to the immutable effects of chaos. As the modern theory of chaos was born, a belief in the determinability of weather forecasts died that day in Professor Lorenz's office. Until then, even the great mathematician John Von Neumann believed that an accurate deterministic forecast could be made indefinitely into the future. All that would be needed to achieve perfect forecasts, he thought, would be better models and better data.

Lorenz showed that this was not possible in a fundamental and mathematical way. Unless the initial data fields were absolutely perfect, no computer model, no matter how sophisticated or how much data could be accessed, would ever be able to forecast the weather perfectly. In principle, because each bit of data possesses a range of uncertainty, an infinite number of possible predictions are possible for a given set of initial conditions within the range of model, measurement, and computational errors. Realization of this principle arrived like a thunderclap in the field of weather forecasting. Meteorology had entered a realm of the uncertain, long inhabited by quantum physicists.

Still, some attempts have been made to partly overcome the debilitating effects of chaos on weather forecasts. The model-of-the-day method, previously discussed, is a somewhat different and much more simple-minded form of consensus forecast, one that requires less computer time but more judgment on the part of the human forecaster. A recent innovation in making weather predictions using complex mathematical models is the so-called ensemble forecast. In recognition that the initial map conditions of pressure, temperature, wind, and humidity are subject to

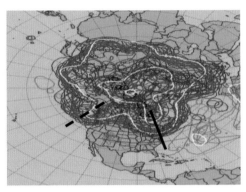

FIGURE 5.6. A spaghetti plot of the 500-hPa flow across the Northern Hemisphere (consisting of 20 different forecasts from differing initial conditions), 10 days ahead during midsummer. Red, green, and blue contours denote the 576-, 552-, and 528-dm height contours, respectively. The white contours are the 576- and 552-dm contours as generated from the conventional high-resolution operational forecast. The black dashed and solid lines, respectively, denote areas where the location of the trough axis is uncertain (a high degree of scatter) and where the location of the trough axis exhibits a high degree of confidence due to the coherence in the contours. Courtesy of Penn State's e-Wall.

error, ensemble forecasts are made from one initial starting time, but with somewhat differing initial map conditions for each of a number of forecasts. Ensemble forecasts constitute another method of employing consensus. The resulting set of forecasts is then blended together to produce a consensus forecast, which turns out to be superior to any one of the individual forecasts. Moreover, as in the model-of-the-day method, areas in which the ensemble forecasts show a lack of agreement between the various forecasts, highlight where the consensus forecast might be less certain of a correct solution.

One such ensemble forecast is the so-called spaghetti plot, an example of which is shown in Figure 5.6. A spaghetti plot consists of a multitude of forecasts for the 500-mb pressure level (representing the isobars at approximately 5500 m [550 dm] above sea level), starting from a large number of slightly different initial conditions. In this case, the red, green, and blue lines, respectively, correspond to the 576-, 552-, and 528-dm contours. The white lines pertain to the 576- and 552-dm contours generated from the conventional high-resolution forecast. Clearly, there is some discrepancy between

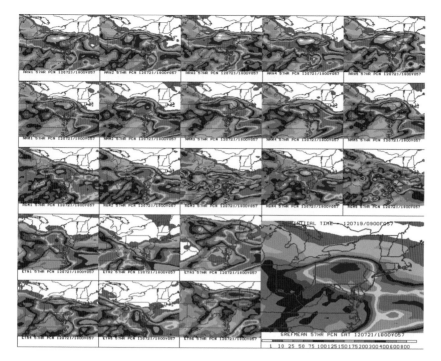

FIGURE 5.7. A display of the 21 predictions including the consensus forecast (lower right) based on the Short Range Ensemble Forecasting (SREF) system of the accumulated rainfall 57 hours in the future over the Northeast. Courtesy of Penn State's e-Wall.

the operational and the spaghetti plots as to the locations of troughs and ridges, as well as considerable scatter in parts of the map.

Areas of wide diversity between the contour lines derived from these ensemble forecasts, such as in locations of troughs and ridges, constitute areas of uncertainty in that part of the flow pattern (e.g., denoted by the black dashed line in the trough west of North America in Figure 5.6). Areas where the contours are very similar indicate means that the ensemble solutions more or less agree (the black solid line in the figure), suggesting a high degree of confidence in the location of that feature—in this latter case a trough over eastern North America. Not shown is the consensus contours for the spaghetti plot; these would most likely have been the optimum forecast.

The reasoning behind ensemble forecasting is relatively straightforward: (1) There is uncertainty in our observations; (2) there is uncertainty in the analysis; and (3) there is uncertainty in the physics that we model. The various ensembles are a logical approach to minimize the uncertainties in the forecasts in a statistical and visual manner.

Another version of the ensemble forecast is shown in Figure 5.7. Here, 21 different forecasts of precipitation were made with differing initial conditions, resulting in 21 different forecasts for the rainfall area centered over New York and Pennsylvania, each differing slightly from the others. The slightly enlarged analysis at the lower right-hand part of the figure is the consensus forecast, produced by averaging the 21 forecasts together, presumably a more reliable and accurate prediction of the rain area and amounts.

How computer models can predict the weather

Von Neumann was not the only scientist who thought during the 1940s that weather forecasting was possible using the Newtonian laws of motion for a fluid. Von Neumann, despite his brilliant mind, thought erroneously that perfect forecasts can be made using the laws of mathematics and physics. As we have just mentioned, Lorenz was able to demonstrate that this was not even mathematically possible.

After all, reasoned Van Neumann, the Newtonian laws of motion are deterministic and as such contain time as a variable. In principle, therefore, it is possible to solve these equations for some time in the future, as Von Neumann had thought possible. In reality, however, the equations are too complex to be solved exactly, so they must be solved inexactly in what is known as numerical weather prediction. Moreover, the equations must be solved at an array (or matrix) of fixed grid points in the horizontal and vertical, rather than at the points where the measurement of temperature, pressure, and humidity were made. An infinite number of such grid points obviously would be impractical. Clearly, the establishment of a finite grid mesh constitutes just another source of error, fundamentally limiting accuracy in weather prediction using computer models. In solving these equations using inexact numerical treatment of the equations, given that time is a variable, it is still possible to solve for the solution a very short interval into the future, although the solution can be exact only when solved for an infinitely short time interval into the future. Clearly, solving the equations for the weather an instant into the future is useless and would certainly take far more time to execute than the instant of time in the equations.

A compromise with the numerical solution to these Newtonian equations would be to solve for the temperature, pressure, wind velocity, and so forth a half-hour or an hour into the future. Besides incurring yet a bit more numerical error in the prediction by extending the time interval well beyond an instant, the prediction of these variables such a short time ahead would still not be very useful.

Instead, a forecaster makes a prediction on the computer using the mathematical model, say, one hour into the future, and then reconstitutes the initial temperature, pressure, wind velocity, and other variables on the weather map for a time one hour into the future, using the predicted values of these parameters. Then, the forecast is restarted from that new map with its new set of variables, starting one hour into the future, making another one-hour forecast. By reconstituting the map with these newly predicted values of pressure, temperature, humidity, etc., the process is repeated and repeated, extending the prediction, leapfrogging in one-hour steps into the future while continually reconstituting the fields of temperature, pressure, and humidity. In this way the forecast marches by steps ahead at hourly intervals, continuing to update the newly predicted variables every hour with the successively predicted ones, until the forecast is potentially extended days into the future.

Of course, the inexactness of the solution—not to mention the use of a finite grid, the slight errors in the initial data, and the empirical nature of the model's parameters (such as the soil moisture, surface reflectance, surface roughness, or vegetation cover)—causes the errors in the forecast to build up with time and eventually diverge so far from what would be the correct forecast that the solutions to the predicted equations become useless. Nevertheless, as we have mentioned, forecasts today have considerable skill out to 5–7 days and often as far as 10 days, although the quality of these forecasts decreases with time from the initial forecast, sometimes gradually and (unfortunately) sometimes very quickly.

Fifty years ago such forecasts were not possible. It is with the vastly superior computer power, only available since the 1980s/90s, that such extended forecasts are possible on a reasonable time scale for execution (e.g., an hour or two to execute and prepare). Lorenz might have had the sobering thought that the inherent errors in the mathematical solutions, not to mention an irreducible level of inaccuracy in the data, would always prevent weather forecasts from being perfect and would likely restrict utility much beyond a week or two. Yet, it is also somewhat remarkable that the steady improvements in the quality of the computer forecasts over the past decades have been due, in part, to increases in the resolution of the computational grid, even though the number, density, and quality of the observations have changed very little.

Figures 5.5, 5.6, and 5.7 reflect both the strength and the weakness of extended forecasts. Although the skill of these forecasts is far beyond that of an individual forecaster working without the aid of computer-generated models, the forecaster often can match or exceed the skill of the computer products for short-term (0–36 or 48 hours) predictions using his or her own eyes and intuition, although perhaps aided by the computer models. Even for longer periods, the modern forecaster can (so to speak) stand on the shoulders of these models and see beyond their capabilities. We will emphasize the role of the human observer in the next chapter.

Forecasts beyond a week or two: laptop solutions

Depending on one's degree of computer savvy, there are numerous ways to expand the forecast horizon with a laptop computer and the Internet. A method of prediction that has a modest degree of success in longer-range outlooks is called an analog forecast. Although we pointed out in Chapter 4 that analog forecasting proved to be unsuccessful for daily forecasts when initiated back in the 1950s, some quite different modern techniques have shown promise for predicting long-term anomalies of temperature and precipitation.

While not based on the physical processes that drive atmospheric motions, this method supposes that current deviations from normal temperature and moisture patterns (called anomalies) can be matched with similar anomalies in the historical record to determine what may happen weeks and even months in the future. The methods of analog forecasting are as varied as the forecasters, but there are some basic principles to keep in mind:

- The larger in area, length of time, and magnitude of an anomaly, the better it will be as an analog predictor.
- Mixing and matching temperature, precipitation, and drought index anomalies may provide a better physical reasoning for an analog.
- Most analogs are only effective in the 1–6 month period.
- Analogs that account for the modulation of longer-term seasonal indices may be more robust over time and somewhat better predictors for more than one month ahead.

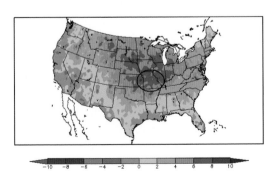

FIGURE 5.8. Departure from normal temperature (°C) over the United States during July 2012, the maximum being in excess of 6°C (circled area). Courtesy the NOAA Central Library Data Imaging Project.

Figure 5.8 shows the temperature anomaly for the United States during July 2012. The most severe warmth, amounting to a departure from normal of over 6°C (accompanying a severe precipitation deficit—not shown), was centered over the central plains (circled). Analog forecasts following a sequence of rules based on previous anomalously warm patterns in that region show that such a pattern of warmth would persist for just another two months and then vanish in October. By contrast, the precipitation anomaly is predicted to disappear by September.

In the next chapter we will discuss how an amateur observer can predict the weather using just a few simple rules, combined with knowledge of the weather map and cloud forms discussed in earlier chapters.

CHAPTER 6

THE OBSERVER'S GUIDE TO WEATHER FORECASTING

Using one's eyes to forecast the weather

A goal of this book is to increase the reader's powers of observing the sky and interpreting the significance of the evolving cloud drama. One's eyes and intuition can help improve weather forecasts for periods of up to a day or two into the future, even without the aid of a computer. There are many occasions when a correct interpretation of the sky can help the forecaster as well as the interested amateur. In one sense, observing the sky has become almost a lost art in operational weather forecasting with the increasing availability of seductive machine-generated products. On the other hand, one's eyes are sometimes superior to such products for short-term prediction, especially in the so-called (and aforementioned) nowcasting problem, where local forecasts in rapidly changing weather conditions are required of the professional weather forecaster. As we pointed out in Chapter 4, local microclimates are embedded within the larger regional weather patterns and, within these local microclimates, smaller-scale microclimates are to be found—and so on down to the smallest scales. For this reason, the local weather forecaster often can make better predictions for his or her town than can the National Weather Service forecaster whose job is to make a regional forecast.

Some popular books on weather simplify the problem of prediction for the amateur weather observer by including a table, in which the columns and rows specify two parameters (e.g., wind direction and barometric pressure), from which the weather forecast is obtained based on knowledge of both of these parameters. One chooses the current pressure and wind direction and reads a forecast from the table. This approach is simplistic and somewhat limiting, as more than two or three parameters must be assessed by the forecaster and blended with intuition gained from experience to arrive at the best possible prediction.

For the amateur weather forecaster who may not have access to or interest in the various machine-generated products, the eye and a few simple measurements are essential for understanding the current weather and for making short-range predictions. The following discussion emphasizes the role of observations that can be made by the amateur and professional forecaster alike.

The simple but essential tools available for good observations are the wind direction, state of the sky, changes in sky cover, cloud formation, cloud type, barometric pressure, pressure tendency, knowledge of the classic cyclone model and its attendant variations, and intuition.

A good way to teach forecasting is by examples in which all of the above factors are taken into consideration. Accordingly, we next present some highly representative scenarios that might be experienced by an observer but based on the principles discussed in Chapters 1 and 2 with reference to cloud types and examples presented in Chapter 3.

Scenario A: The observer remains north of the warm front during passage of a classic comma-cloud disturbance

Consider the approach of a low pressure disturbance. What indicators would one expect to observe as this scenario unfolds? To show the progression of cloud forms and weather during the passage of a low pressure disturbance with its attendant comma cloud, we imagine the passage of such a weather system using the classic cloud pattern, previously introduced in Figures 2.3 and 3.7 and now reproduced here as Figure 6.1. Because it is not possible to depict the movement of a stationary drawing on a page, we visualize the observer experiencing a sequence of weather events as if he or she were transported from right to left along the paths labeled A or B (similar to scenarios A and B in Chapter 3) in Figure 6.1, simulating the real case in which the cyclone moves past the observer. This presumes the passage of the weather pattern without change of shape or intensity, so that the observer experiences the same series of events that are shown from east to west along either of these two paths. As such, the trace does not signify any specific geographi-

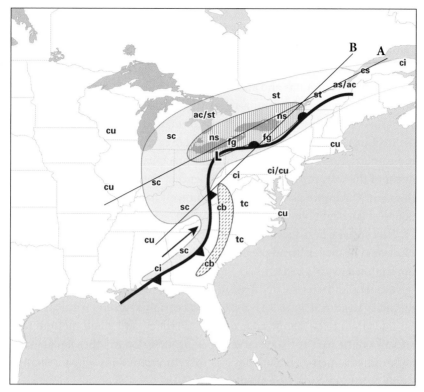

FIGURE 6.1. Cloud pattern and cloud types associated with a classic comma-cloud pattern (identical to Figure 3.7), but with traces labeled A and B denoting the sequence of cloud cover and cloud types experienced by an observer initially at points A (scenario A) or B (scenario B).

cal location or direction that the weather pattern moves, except insofar as it is moving relative to the observer. The system might be moving from any direction, including northward along the Atlantic coastline as a coastal storm or through an Alberta clipper disturbance (discussed in Chapter 4), which moves toward the southeast from western Canada. For the purpose of discussion we will assume that north is directed toward the top of Figure 6.1.

The reader is encouraged to look at the first several figures in Chapter 2 (the comma-cloud model), and the cloud forms shown and scenarios described in Chapter 3 to become oriented with respect to the pressure, temperature, and sky cover changes that would be experienced in the classic comma-cloud pattern.

Let us first consider, in turn, the clouds, weather, wind direction, and barometric pressure experienced by an observer in scenario A in Figure 6.1.

Clouds and weather

As discussed in Chapters 2 and 3, the sky cover observed during the approach and arrival of a classic low pressure system is a progression of cloud types and cloud amounts, starting with wispy cirrus, thickening to cirrostratus, lowering to altocumulus and altostratus, then to stratus, and finally to nimbostratus as precipitation begins to fall (the first part of the segment labeled scenario A in Figure 6.1). Although clouds may advance from a wide range of directions, the thickest part of the cloud cover will appear to be emanating more or less from a central azimuth. If the precipitation area, typically found north of the warm front, is to be experienced, the clouds will thicken with this central azimuth anywhere from southwest to perhaps slightly north of due west (the light blue sector in Figure 6.2).

This section best serving as the harbinger of an approaching storm is illustrated in the left-hand pie chart in Figure 6.2. The weather, however, is not confined to 90° quadrants. The reader should bear in mind that the exact quadrant boundaries are not fixed in stone with regard to their implications for future weather. Figure 6.2 is meant to serve only as an empirical guideline for interpreting the significance of wind direction and cloud motion. Discretion on the part of the observer—paying attention to other factors such as barometric pressure tendency—is therefore called for when assigning significance to a particular quadrant.

Clouds thickening with a central azimuth from the southwest (the yellow sector) are likely to be harbingers of an approaching disturbance in which precipitation can be expected. Clouds thickening from the northwest (light

important cloud sectors   important wind sectors

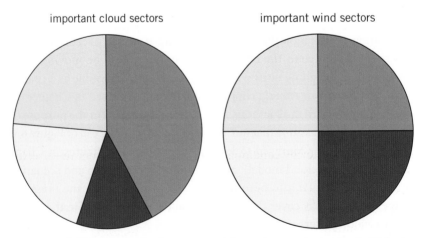

FIGURE 6.2. (left) Important cloud sectors pertaining to the central azimuth of a thickening cloud shield; (right) same but for the wind direction.

blue sector) are likely to presage situations where the storm is moving north of the observer or where a weak clipper disturbance is approaching. Clouds thickening from the south (maroon sector) are likely to indicate that the storm will pass to the south of the observer or that, for those living near the Atlantic seaboard, the observer may experience the western fringe of a coastal storm. The purple sector represents directions from which clouds appear to thicken as being highly unlikely to presage any significant disturbance; they may simply be the residue of a disturbance that has already passed the observer.

As stratus give way to nimbostratus, preceded by virga from the stratus (an indicator of incipient precipitation produced by shafts of falling rain or snow), precipitation will be more or less continuous within the cross-hatched area in Figure 6.1.

Winds

In scenario A, the winds will tend to remain from a southerly direction. Winds coming from the southwest and southeast (the yellow and maroon sectors in Figure 6.21) presage the arrival of a disturbance, provided that both the aforementioned cloud progression and the following wind criteria are met. In winter storms, southwesterly and southerly winds (the yellow quadrant) almost always accompany warm-air advection and will usually precede rain or freezing rain rather than snow, though the precipitation may start as snow. Snow is likely to be accompanied by winds from the northeast or east (the purple quadrant), directions that accompany cold-air advection and therefore presage falling temperatures. Winds from the southeast (the maroon quadrant) indicate that the observer is situated, north of the warm front. Winds turning from southeast to southwest also may indicate the passage of a warm front, as indicated in scenario B.

In winter, winds from the southwest presage a warming trend and precipitation within 12–24 hours; in summer, that can take 24–72 hours, because weather systems move much slower during the warmer seasons. Winds from the southeast sector (the maroon sector) presage a warming trend and somewhat more stormy conditions than winds from the southwest sector. In winter, winds from this quadrant can mean rain or snow, with the latter changing to freezing precipitation (freezing rain or sleet).

Winds coming from the northwest sector (light blue) indicate fair weather; indeed, we call this the fair-weather quadrant. A change of wind direction to the west or northwest from a southerly direction during a storm signals the imminent end of the storm. Except when lake effects (Chapter 4) are expected or the observer is located within the comma head west of the low pressure center, the northwest sector presages fair weather.

Winds from the north or northeast might, in some circumstances, accompany stormy weather, but this direction almost always signals cold-air advection. Nor'easters often are accompanied by winds from the north or northeast, as the name implies.

Winds from the east or northeast sector (purple) can indicate the approach of a storm from the south, a coastal storm, followed by colder weather. In general, the stronger the winds, the more rapid the pressure changes with time.

Pressure and pressure tendency
The approach of a disturbance will be preceded by falling pressures—rapidly falling pressures in the case of the rapid approach of an intense low pressure system, such as a coastal storm. The end of the storm is signaled by a change in wind direction to northwest and a rising pressure. Sharply rising pressures indicate the passage of the cold front or that the storm center is moving rapidly to the east. Pressures below 29.60 inches of mercury (1002 mb) indicate that precipitation is imminent, if it has not already begun. Sea level pressures above 30.40 inches (1030 mb) indicate that precipitation is likely 24 hours or more away—even longer in summertime.

The end of the storm
Once the storm has passed, a very low pressure (well below 29.60 inches; 1002 mb) means that the weather may take up to a day or more to clear, even after the winds have changed to the northwest. Sustained precipitation quickly ceases after the wind changes direction and pressures begin to rise. Nimbostratus will give way to stratocumulus or stratus formed by the spreading out of stratocumulus, and by the return to northwest or westerly winds. These types of clouds can be differentiated from layer clouds associated with large-scale ascending motion and sustained precipitation. As mentioned earlier in this book, the clue that a stratiform deck does not indicate storminess is that clear, blue sky is visible through holes in the overcast. Stratocumulus will tend to persist until the isobars are no longer curved cyclonically (enclosing lower pressure) but become straight or curved anticyclonically (enclosing higher pressure). In the former instance, frictional drag will require the surface winds to converge while, in the latter case, frictional drag will impel the winds to diverge.

As the storm matures, the dry tongue becomes increasingly prominent and tends to move close to the center of the low. Once this happens, the low-pressure system begins to weaken and may separate itself from the warm front. Thus, the end of the storm may coincide with a brief clearing as the dry tongue passes, accompanied by a strong northwesterly or westerly wind followed by a return to stratocumulus overcast and the possibility of intermittent snow or rain squalls. Eventually, the clouds break and are replaced by fair-weather cumulus west of the comma head.

Scenario B: warm front and warm-sector weather

Many winter storms approach the northeastern part of the United States from the Ohio River valley and move north of the observer. While an infinite number of possible scenarios exist, the one labeled scenario B in Figure 6.2 is highly representative of many of these storms. In general, these cyclones are not coastal storms and seldom experience explosive cyclogenesis, although they may deposit a mixture of snow, sleet, freezing rain, and rain as they pass. Scenario B represents a situation where the observer experiences the passage of the warm front, thereby entering the warm sector and later the passage of the cold front, or the observer remains slightly north of the warm front.

Clouds and weather
While initially the progression of cloud cover may resemble that for scenario A, which showed the successive stages presaging the arrival of a storm, the warm-front passage coincides with the cessation of this progression, to be replaced by a mixture of fair-weather clouds and a chaotic array of cirrus and altocumulus within the warm sector. In the instance where the observer remains slightly north of the warm front, precipitation is usually in the form of rain or freezing rain in winter, accompanied by fog.

Within the warm sector, the chaotic cloud pattern tends to thicken with the immediate approach of the cold front, becoming a uniform cover during and just after the arrival and passage of the front. Particularly in spring and summer, the approach of the cold front may be accompanied by towering cumulus presaging the arrival of a thunderstorm squall line. With the passage of the front and some brief showers with perhaps lingering stratocumulus, fair-weather clouds (cumulus) return. In winter, the passage of a trailing cold front south of a deep low pressure cell often engenders very severe convective storms, accompanied by tornadoes, hail, and destructive winds. These outbreaks are particularly frequent in the southeastern states during the winter and early spring months.

In situations where the classic sky precursors to the arrival of precipitation lead to precipitation immediately north of the warm front, low nimbostratus, accompanied by sleet or freezing rain and fog in the winter, will give way to partial clearing if the warm front passes.

Winds
Wind directions will be much the same as in scenario A, ranging from south-west to southeast, but will become more southerly or southwesterly once the warm front has passed. Later, the wind direction will swing to the west and northwest with the passage of the cold front.

Pressure tendency
As in scenario A, pressure will fall steadily until the warm front passes, after which the pressure may briefly rise or will remain fairly steady or decrease slowly in the warm sector until the cold front approaches. At that point, pressures will fall rapidly for a short period and rise abruptly with the passage of the cold front.

Other scenarios
We could go on and create a vast and untenable number of such scenarios. The readers are welcome to apply their understanding of the principles discussed in this book, including cloud forms, cloud progression, pressure and pressure tendency, and wind direction. Imagine how these observations might change in other scenarios. For example, what would these observations look like if the track of the disturbance took the observer on a path just south of the track of scenario B, whereby the observer remains in the warm sector the entire time until the passage of the cold front? Or, a track north of that in scenario A?

Once the principles discussed in this book are understood, it is worth pointing out that the weather map and its attendant cloud and precipitation patterns may deviate greatly from the classic model. For example, Figure 6.3 shows a rather complicated (but not infrequently observed) pattern of fronts and precipitation, which do not closely adhere to the many examples we have presented demonstrating the classic comma-cloud pattern.

Rain versus snow: how to forecast

An urgent and frequent demand on forecasters during the winter is to decide whether the precipitation will fall as snow, rain, or some intermediate type of hydrometeor, such as freezing rain (rain freezing on contact with the surface), hail (rain freezing as it nears the ground), or sleet (snowflakes melting then refreezing near the surface). Because of the irregularities of the

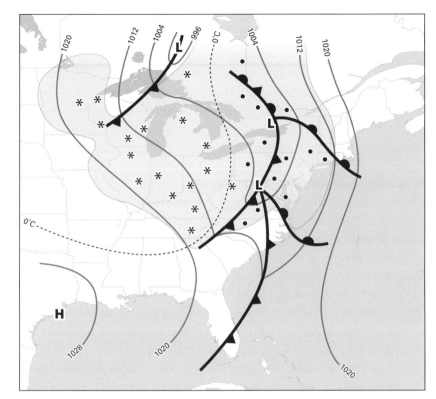

FIGURE 6.3. Surface weather map for December 1, 2010 at 7:00 am EST showing sea level isobars labeled in mb, centers of highs and lows, surface fronts, and the area of precipitation (shaded). The thin dashed line represents the 0°C isotherm at the pressure level 850 mb (approximately 1,700 m above sea level), and the symbols * and • denote surface observations of snow or rain, respectively.

terrain (e.g., pooling of subfreezing air in valleys), the prediction of the 0°C isotherm at the surface may be difficult to specify because of the topography. Even if the forecast of the surface freezing isotherm were perfect, its location might not be indicative of whether the precipitation will fall as liquid, frozen, or something in between.

One fairly reliable diagnostic of the rain–snow line is the 0°C isotherm at the 850-mb pressure level, approximately 1,700 m above sea level. Figure 6.3 shows the distribution of rain and snow for the complex disturbance referred to earlier. Snow and rain are neatly separated by the 0°C isotherm, which is much more easily and confidently predictable using the computer model output than the freezing line on the surface. Moreover, the latter is still a useful field to plot, as regions where the surface temperature is below freezing but above freezing at 850 mb are likely to correspond to situations in which freezing rain, hail, or sleet might occur. Conversely, where the temperatures at 850 mb are below freezing but above freezing at the surface, there will likely be precipitation in the form of wet snow. The latter situation frequently occurs in the late fall and can prove highly destructive of vegetation when wet snow is intercepted by leaves still remaining on deciduous trees. (Websites for analyses and predictions of such fields as the 850 mb temperature are provided in the Appendix.)

Diagnosing cold versus warm advection

Chapter 2 emphasizes the importance of temperature advection in its relationship to rising or sinking motion along sloping surfaces (large-scale slant ascent or descent). The type of advection (warm or cold) and perhaps rough magnitude (strong or weak) of the temperature advection can be inferred from observing the direction of the wind at two levels in the atmosphere, the surface and at middle or high levels. While the following diagnosis rests on a firm physical and mathematical basis in which one can actually compute a close approximation to the magnitude of the temperature advection, the procedure outlined here is qualitative and quite simple.

Here is the procedure: First, determine the wind directions at two levels, the surface and at a level representative of the motion of middle clouds (such as altocumulus) or high clouds (such as cirrus). Next, plot these directions as arrows from the center point of an imaginary circle in the direction that the wind is going. In the example of Figure 6.4, the surface wind is northwesterly—that is, *from* the northwest—and the wind at some upper level is southwesterly—that is, from the southwest.

The next step is to determine the smaller of the two possible angles between the surface and upper-level wind directions. By smallest angle, we mean an angle that is less than 180°. In this case, the smaller angle is about 70°, whereas the complimentary angle, 290°, is the largest angle and so the larger angle would be incorrect for applying this method.

Finally, mentally rotate the lower-level (surface) wind arrow toward the upper-level wind arrow through this smaller angle as in the figure. In this example that angle is represented by the curved arrow, rotated from lower to upper level.

If, in moving from the surface to the upper levels, the angle of rotation is counterclockwise, as in this example, cold-air advection is occurring. If the rotational direction through the smaller angle is clockwise, warm-air advection is occurring.

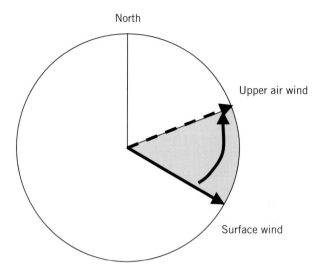

Wind direction and temperature advection

North

Upper air wind

Surface wind

FIGURE 6.4. Wind directions at the surface (solid, straight arrow) and at middle or upper levels (dashed arrow) and the rotation angle from surface to upper air (curved arrow) for a hypothetical case where cold advection is occurring.

The strongest advection occurs when the angle is closest to 90°. Advection is zero when the angle is either 180° or 0°. The stronger the winds (and the greater the angle up to 90°), the stronger the advection—whether warm or cold.

As we will discuss in further detail below, in conjunction with our discussion of how to take observations, surface wind direction can be easily obtained, either by noting the direction of a weathervane, smoke emerging from a stack, the direction of the flag waving from a high pole, or the low cloud motion. Cumulus clouds often form themselves into rows consisting of elongated, cigar-shaped clouds oriented parallel to the wind direction with the more ragged ends downwind. Swelling cumulus tend to tilt or even collapse downwind. To determine the wind at the surface, it is more reliable to observe the direction of low clouds or smoke emission from tall smokestacks than from the wind at eye level, which is subject to turbulent eddies that, because of their circular shape, can induce a wind from any direction of the compass as they pass the observer. Upper-air winds are more difficult to determine, unless middle- or upper-level clouds are present and one has the patience to observe their motion. Quite often, middle- and upper-level clouds are oriented along the direction of the wind, forming rows (as with cumulus clouds) or streaks.

Learning which details are the most important

It is important to remember that a number of the idealized views of weather systems presented in this book will need to be tweaked by experience. The easiest of these to conceptualize is the Norwegian Cyclone Model, manifested visually as the comma-cloud pattern. While the latter remains essentially valid, almost a century after its conceptualization by Scandinavian meteorologists during World War I, much of the theory has been updated since then.

Yet it is rare to find a weather chart or satellite image that matches this so-called perfect depiction of a storm. As we have already seen, no two storm systems are exactly identical and the differences, albeit subtle on some days, are due to changes in the environmental conditions that are usually discernible at the start of the forecast period. This process can also involve changes within an evolving weather pattern. Even the computer can be fooled, requiring the computerized maps to gradually adjust to new conditions, as was illustrated in regard to Figure 5.5. Whatever the specific cause, this discussion highlights that attention to details of the current weather pattern can provide useful information for the forecast process. The challenge is determining which detail holds a key to today's weather pattern.

Another example of the importance of details in the observation can be noted with a large slow-moving storm. Areas to the north of the low center and the storm's track can remain cloudy and chilly for several days, but the character of the day's weather can change substantially, and the surface weather maps shown in Figures 6.5 and 6.6 illustrate this point. These figures show a classic comma-shaped pattern similar to those presented in Chapters 2 and 3, with warm and cold fronts intersecting and an occluded front closely resembling the example shown in Figure 2.8. Notice the differences in the precipitation pattern along the cold front on these two separate occasions. The clear area behind the cold front in Figure 6.5 differs markedly from the 200-km-wide band of cloud and precipitation just behind the cold front in Figure 6.6. Cold fronts that exhibit little cloud behind them have the technical name katabatic fronts, so named because they are characterized by a strong sinking motion in the colder air just behind the front. Their opposite, anabatic fronts, tend to have a shallower layer of cold air and experience overrunning of the air originating in the warm sector, similar to the process associated with the warm front, discussed in Chapter 2. A warm front, therefore, is a classic example of an anabatic front.

Technically speaking, the difference between a katabatic and an anabatic front means that, in the latter case, the slantwise ascent (which produces

FIGURE 6.5. A katabatic cold front with precipitation preceding the front. Courtesy of NOAA.

FIGURE 6.6. An anabatic cold front with precipitation following the front. Courtesy of NOAA.

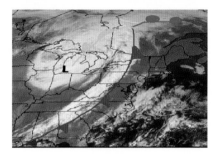

clouds and showers) is also occurring behind the cold front as well as in the cold air ahead of the warm front.

A further interesting feature seen in these figures, and also apparent in Figure 2.8, is the circularity of the comma-head cloud surrounding the occluded low. Despite the highly simplified depiction of the effects of frictional convergence illustrated in Figure 2.14, the figure does suggest that the low-level stratus and stratocumulus clouds induced by shallow frictional convergence tend to form a nearly circular shape corresponding to closed isobars around a low pressure system, albeit that this type of cloud is confined primarily to the area west of the storm center. Most of this precipitation in this frictionally induced stratocumulus cloud would consist of light showers or snow squalls, with the more sustained precipitation occurring farther to the northeast, where large-scale slant ascent is occurring in conjunction with the frictionally induced ascent. As the cold front passes, showers behind the front, whether brief or in a wide band as in Figure 6.6, would give way briefly to clearing skies in the rather prominent dry tongue (Figure 6.6 as illustrated in Figure 2.3), followed by a return to overcast and showery conditions until the cyclonically curved isobars have moved on to the east.

Mesoscale structure of the passing cold front
Perhaps a subtler conceptual model is the mesoscale version of the cold front. Let us consider the cold front, discussed in Chapter 2. Often times, a

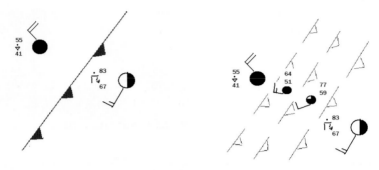

FIGURE 6.7. Two schematic views of the same cold front. The figure at left shows the conventional representation of a cold front on a weather map, including an observing station, initially situated east of the front (right side; prefrontal location) and subsequently west of the front (postfrontal location). Observations consist of temperature (top left number in degrees Fahrenheit), dew point (bottom left number), weather (the symbols between the two numbers—showers for the postfrontal observation; thunderstorms and showers for the prefrontal observation), wind direction, wind speed, and cloud cover (the interior of the station circle—overcast for the postfrontal observation; half overcast for the prefrontal observations). The right-hand side shows the same pair of observation, plus two others made during the passage of the front with successive changes in weather conditions and cloud cover.

cold front is thought of as a single boundary separating airstreams of differing densities and differing temperature gradients. On a small scale, the cold front typically comes in several surges over a period of up to several hours or more, with successive pushes of denser, colder air causing successive wind shifts and pressure rises. Sometimes, weather maps will depict two or more cold fronts separated by some distance, each front constituting a further surge of cold air and successive wind changes toward the northwest.

The visualization in Figure 6.7 shows this contrast. With the passage of the front, temperature, dewpoint, weather, wind direction, and speed change quickly while requiring some time before the wind has turned to the northwest and the weather has established itself as a typically cold, postfrontal, showery regime.

A main goal in forecasting is to allow one's experience, especially with classic weather systems such as fronts and large-scale storms, to be tweaked so as to have a more realistic expectation of the next event.

Making observations and a forecast

Good observations—paying close attention to the evolving sky panorama—can help both the professional forecaster and the amateur to better understand the meteorological events taking place and thereby to improve their forecasts, especially those short-range, local forecasts referred to as now-casting. For the amateur, keeping a weather diary is a very useful method for both making and interpreting the observations and for learning from previous weather events. Table 6.1 provides a sample template for recording observations, measurements, and comments. The template can of course be modified to include whatever information or data the forecaster feels necessary to improve the process, or even to revise the format for the log. The sample observations here might be representative of an approaching disturbance.

To some extent, a simple table such as this may be too restrictive to include all of the observer's impressions, so one might prefer to fill an entire page with a single day's observations. With the vast storage capacity of today's computers, such an extensive log could easily be managed in a formal database.

Table 6.1 contains a number of columns, starting with date and time of day. Next, cloud amount and type are recorded. In this case, the observer notes some thin cirrus covering the western part of the sky. Without a history, the presence of cirrus does not in itself signify the approach of a low-pressure disturbance, but the observer also notes in the remarks that the clouds have been moving and thickening from the west since dawn. This illustrates that it is not only important to point out the instantaneous state of the sky, which is only half the story, but also to note the progression of cloud types present or not present during the previous several hours or more. For example, the weather might be stormy, in which case one might wish to note the time of onset of precipitation, its intensity, and perhaps when a change in phase occurred (e.g., snow to rain).

Temperature is another useful measurement, as is the dewpoint. Ideally, an outdoor max–min thermometer would be helpful. Dewpoint can

TABLE 6.1. Template for a daily weather log.

Date	Local time	Clouds (amount, type, motion)	Weather	Temp	Dew pt.	Wind dir; Speed	Bar/ Tend.	Precip	Remarks/ F'cast
1/12/12	0900	Thin ci in western half of sky. Clouds moving from west since dawn.	sunny	−1C	−8C	SW light	30.16 in ↓ slowly	0.0 in	Warm air advection ci thickening in W; since dawn. F'cast: Rain by next day.

provide additional information. For example, we showed in Figure 6.3 that the relationship between surface and 850-mb temperatures is important for resolving the issue of rain versus snow. Dewpoint is also an essential measurement when assessing the likelihood of thunderstorms, including severe thunderstorms. (Recall our dewpoint criteria for thunderstorms and severe thunderstorms mentioned in Chapter 3.) Dewpoint, being relatively constant during the day, also serves as a reasonable predictor of low temperature for the coming night, as air temperatures tend not to decrease at night below the previous afternoon's dewpoint.

In situating the temperature sensor, it is important to keep it out of the direct sun and away from nearby objects that can become directly heated by the sun, such as the side of a house exposed to the sun. A sensor placed in the shade may still be affected by the sun if, for example, it is placed in the shade under an eve or roof overhang but above the side of the house exposed to direct sunlight. An optimally sited instrument might require some close inspection of the patterns of sunlight and shade during the day and at different times of the year. Moreover, it is important to situate the temperature sensor not too close to the ground but where it is well ventilated by the air, while still being easy to read, ideally about 1.5 m above the surface.

The same caution must be exercised for the dewpoint sensor. Conventional dewpoint thermometers, called psychrometers, are somewhat cumbersome to use and interpret, as they require a set of tables to convert the psychrometric reading, or wet bulb temperature, to dewpoint. Handheld psychrometers also require some vigorous arm movements to ventilate the wetted bulb. We recommend that both the temperature and dewpoint be measured using an inexpensive weather station, such as can be purchased for under $75. These devices, which also measure atmospheric pressure and wind speed and direction, transmit measurements remotely from the outdoor sensors to a display panel that can be located in one's kitchen or living room.

Wind direction is an essential observation, as has been emphasized in previous chapters and reiterated in regard to Figure 6.2. One need not mount a wind vane atop a house in order to make this measurement. Wind direction can be assessed looking at the movement of low clouds, smoke rising from chimneys, a waving flag, or the direction of falling leaves or bits of grass tossed into the air. The sign of the temperature advection can be extracted from these measurements, using the method outlined in Figure 6.4. Since the surface wind recorded in Table 6.1 is from the southwest and the middle- and upper-level clouds seem to be moving, as well as advancing, from the west, the rotation of the wind direction from surface to upper levels is clockwise, indicating warm-air advection.

Wind speed is not as important as wind direction, but it should be noted qualitatively, as it serves as a measure of how closely spaced the surface isobars are. It is sufficient to record the wind speed in just three or four categories:

- Calm or nearly calm (smoke rises straight up; flags hang motionless)
- Light breeze (smoke has a definite direction; leaves on the tops of tall trees rustle slightly)
- Moderate breeze (tops of trees sway a bit; wind is felt in one's face at street level, smoke blows almost sideways from chimneys)
- Strong wind (trees sway, tree limbs or wires start to whistle; dust blows around; smoke is ripped into fragments upon leaving the chimney)

In many respects, observations of wind speed and wind direction derived from low cloud motion are more representative than those measured using a wind vane and an anemometer situated atop a house or garage, because the latter is highly subject to distortion due to turbulent eddy motion created by surrounding trees, houses, and other obstructions. Such eddies as they pass by the observer can cause the surface wind to rotate 360° in just a few minutes. Moreover, because of surface friction, wind speed increases logarithmically with height from essentially zero at grass-top level to something still less than the geostrophic wind speed at about 50 m above the surface, above which the wind speed gradually increases with height and becomes approximately geostrophic several hundred meters above the surface. For that reason, the wind speed often appears calm or nearly calm at levels sensed by someone on the street or by an anemometer at housetop level, even though leaves are observed to be moving at the tops of tall trees. At night, because of the presence of a nocturnal inversion and its suppressive effect on turbulent exchanges of air between the surface layer and that farther aloft, wind speeds decrease toward zero and the air may become still at all levels below a few hundred meters above the ground. Accordingly, wind direction and speed are best assessed at night by observing the motion of low clouds.

As discussed, barometric tendency is extremely important—even more so than the absolute pressure itself. Ideally, one should note the three-hour pressure tendency, as is reported on weather maps, but this observation may be somewhat inconvenient to make, at least systematically and especially at night. Home weather stations can indicate pressure tendency, but the inexpensive ones are generally incapable of measuring pressure tendency over a short a time period of three hours. Twelve-hour pressure tendency measurements may be quite misleading. Still, it would be useful if a few observations of barometric pressure (and therefore of pressure tendency) could be noted at times during the day when an interesting weather pattern is emerging. Changes in pressure of 0.01 in. or less over three or six hours should be regarded as unchanging.

Pressure itself is a broadly useful parameter. How high or low the barometric reading is provides a rough idea of where the observer is situated with respect to the centers of the high or low.

Rainfall amounts, while of very limited prognostic value, can be very useful information, especially for growing fruit, vegetables, or flowers. In summer, rainfall data can provide a rough idea of the soil moisture content and therefore serve as an indicator of the need to irrigate or water the farm or garden.

While home weather stations have the capability of measuring rainfall amounts, it is simple for one to construct a rain gauge using a glass jar and funnel, as shown in Figure 6.8. To construct this device, take a wide funnel and insert it in a tall jar. Ideally the jar should be graduated in milliliters. The steps in constructing a home rain gauge are as follows:

- Measure the diameter of the funnel, which should be made of metal to prevent it from blowing away during a rainstorm.
- Anchor the jar to the surface; perhaps one could attach it with glue to a solid metal or wooden block to keep the jar from blowing over.
- Situate the gauge as far away from objects as possible, such as trees or houses, and at an elevation well above the ground but easy to access.
- To measure the depth of rainfall, either read the milliliter (ml) scale on the side of the bottle or pour the contents of the bottle, after removing the funnel, into a finely graduated flask, so as to measure the volume in ml. The rainfall depth in centimeters is obtained from evaluation of the following simple equation:

Rainfall depth (cm) = [Volume of rainfall (ml) × 1.27]/Funnel diameter (cm) squared

So, for example, if one measures a volume of 100 ml of rainfall using a 10-cm-diameter funnel, the rainfall depth will be 1.27 cm or 12.7 mm (about a third of an inch). Even if the jar has no gradations on its side, it is useful to monitor the rainfall rate during a storm, without having to constantly pour the bottle's contents into a graduated flask. This could be done by calibrat-

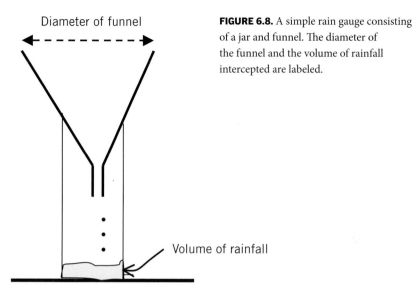

Diameter of funnel

FIGURE 6.8. A simple rain gauge consisting of a jar and funnel. The diameter of the funnel and the volume of rainfall intercepted are labeled.

Volume of rainfall

ing the depth of rainfall in the bottle to the depth of the rainfall by pouring known volumes of water into the bottle and then measuring that with the aid of the graduated flask. Given the diameter of the funnel, one can calculate from these known volumes the rainfall depth and then etch these levels into the side of the bottle or draw a scale on some hardy, waterproof tape.

If a funnel is unavailable, then a wide container such as a coffee would suffice, the diameter for the rain depth formula being that of the can itself. The jar or the funnel should be as wide as possible to maximize accuracy. During its maiden voyage as a rain gauge, the funnel should be observed to make sure that the rain is able to fall into it unobstructed.

Finally, the observer should note relevant observations or ideas, as indicated in the last column of Table 6.1. Repeated observations, whether using this kind of table or a free-form diary, will in time build the observer's understanding of the weather, as successive weather events will eventually create a subconscious library of cases to serve as analog conditions for later forecasts.

Although many of the observations or measurements listed in the above discussion are easily obtainable without any special equipment other than the observer's acuity, a barometer and thermometer (or a simple weather station) are very useful. Some items can be derived from readily available computer products, such as the 850-mb intial and forecasted temperature maps (useful in assessing the rain–snow boundary) and many other products. We have provided a list of possible weather sites and an idea of their content in the Appendix; please note that URLs and products are subject to change.

Takeaway principles: advice to the novice forecaster

Finally, we must emphasize that observations can appear ambiguous to the forecaster. For example, the wind may be coming from the northwest, but the barometer is falling and the sky is almost totally obscured by dense low clouds, but otherwise exhibits no upper- or middle-level clouds. With such seemingly conflicting signals, the forecaster must rely on intuition, perhaps to conclude, despite the falling pressure and the seemingly threatening sky, that the absence of high cloud and the presence of the northwesterly wind indicate continued fair weather and that the low cloud is merely fair-weather stratocumulus. In winter, however, this scenario might signal the possibility of lake-effect snow showers. In other words, all types of observations, including knowledge of the season and of local peculiarities, must be taken into account and integrated with the forecaster's intuition and knowledge.

Three principles underlie good observations: observe the sky with intelligence; learn from what you observe and from your failures in making a forecast; and share what you have learned with others or with yourself via a weather log. In other words, our advice is to become a lifelong learner. Let the habit of observing the sky turn into an opportunity to share experiences with others (through pictures, data networks, and the host of social media outlets).

An important way in which to share one's observations, expand one's interest in the weather, and apply one's knowledge to improving weather prediction is to make new connections. There are numerous ways to take interest in the atmosphere to another level. Depending on how involved the forecaster wishes to get, here are a few examples and ways to make the connection:

- CWOP (Citizen's Weather Observing Program)
- CoCoRaHS (Community Collaborative Rain and Hail Study)
- NWS (National Weather Service) Storm Spotters
- Local TV weather networks
- County Emergency Manager's data network
- Snow network (NWS and/or local TV networks)

Mining data sources

Most readers will probably be content to confine their interest in the weather to making observations and informal weather predictions at home. For those who wish to expand their horizons, a quick Internet search provides an enormous amount of data. In addition to those sources listed in the Appendix, many other atmospheric, oceanic, and environmental databases exist that can be plumbed almost ad infinitum, whereby more connections can be forged, and even more esoteric data can be acquired, such as the Climate Prediction Center, National Climatic Data Center (NCDC), Earth System Research Laboratory, and COMET modules.

The value of timely observations

There is some value to monitoring weather reports frequently, but this can lead to tunnel vision—losing focus on the big picture of what is unfolding in the larger weather pattern. It is a double-edged sword between being a diagnostician versus a prognosticator. Although we have emphasized the value of keeping a log as a way of deepening understanding of the evolution of weather patterns, too much emphasis on understanding the current weather regime can easily swamp a new forecaster in the details, making the prediction too complicated. On the other hand, too little time on analysis of the current conditions can lead even the most experienced predictor to forecast silly things such as a sunny afternoon when it is cloudy outside! Meteorologists, including many meteorology students, sometimes lose themselves in a welter of computer simulations. Sometimes what is missing is to simply read the sky!

So, there is a fulcrum in which prediction must be balanced by analysis and observations, the circumstances dictating where that fulcrum will be. In one sense, it is obvious that the shorter range the prediction, the less time should be spent on data fluctuations, while the longer the forecast period, the more time that would be required both for analysis and prediction.

Generally speaking, the latest data are the best when it comes to shorter-range predictions (one day or less), whereas the most recent observations have little or no bearing on the extended forecast period (3–10 days).

New data streams

The world of weather observations is changing at an accelerated pace, so this discussion is just a snapshot in time of environmental reports that are only dreamed of today but will likely be at our fingertips in years to come. Many new data sources and websites will come into existence, while current ones may change their name or disappear. A list of some current sources include:

- Mobile surface reports (handheld and vehicular)
- Micronets (small-scale, denser observations such as may be made in cities)

- Portable upper air reports
- New satellite data streams

Some "dream" products are already becoming available in the form of current weather reports, radar and satellite images, and a variety of weather maps accessible via mobile devices such as the iPad or iPhone. It is important to remember that having the latest and the best data at one's fingertips is no guarantee of better predictions. In view of the danger of a forecaster becoming inundated by weather data that cannot possibly be digested, it is essential that the forecaster or weather observer determine which data are most useful; sometimes the most useful data are the least available!

Computer simulations

Even more valuable than radar and satellite data, computer simulations and predictions (discussed in Chapter 5) constitute one of the biggest advances in atmospheric sciences during the previous century. While it seems impossible to have a personal satellite or radar, a personal computer is the norm today. In fact, numerical weather prediction, unlike radar and satellites, which require an enormous amount of front-end expenditures, can ride the wave of advances in computer processing units (CPU), storage space, and input–output technology. Indeed, one of the greatest areas of progress in weather prediction during the past few years has been due to greater ability to access and disseminate meteorological data, including those to the public. These computer simulations are currently available to anyone and everyone.

While it is hard to imagine that someday most people will have their own computer forecast model on their iPhone, it is quite possible that applications combined with electronic devices (descendents of the current iPhones) will be developed to downscale and fine-tune forecasts for one's own location, based on the latest version of numerical weather forecasts, such as the Intellicast "app" on the iPad.

Even today, there are opportunities to adjust the computer forecasts to one's own backyard, aside from the model-of-the-day methods outlined in Chapter 5. Many of us already do this. When we hear a forecast low of near 4°C (39°F) in the early autumn, we know which locations within our region may have their first frost because microclimates exist *within* microclimates, such that a certain valley or neighborhood may experience lower temperatures at night than another, or that suburban areas are likely to be cooler than the urban center, or that the temperature at grass level may be several degrees colder than that at eye level on clear mornings. For this reason, no temperature, wind, or other type of forecast can ever be perfectly correct, since these parameters may vary from street to street or even from one side of one's yard to another.

In a sense, one must always try to modify the official forecast and to outguess the forecaster for one's own immediate location. Remember, the best local forecasts for a particular location usually come from local sources, rather from the National Weather Service. This type of personal adjustment of the forecast is especially important in winter in regard to snow amounts. In warmer seasons, one should be able to nudge the expected maximum temperature up a couple of degrees due to the strong spring sun or down a couple degrees at night depending on one's location. Land use and land cover play an essential role in this type of forecast adjustment. For example, thunderstorms often appear to skip over parts of central Pennsylvania where terrain imposes local downslope motions, such as in the lee of the Allegheny Plateau. It can rain heavily in one part of town, while remaining dry in another part of that same town.

There can be a science to these adjustments if one is willing to commit to a fixed method. For example, with temperature predictions, given a properly mounted max–min type thermometer (placed in the shade, about 1.5 m above the ground and away from obvious heat sources), one can keep a record of the local temperatures and compare those with the nearest National Oceanic and Atmospheric Administration's (NOAA) recording station (typically an airport) and also with the predicted maximum and minimum temperature from the computer forecast (using the same "guidance"). After several months (three at a minimum), one should be able to calculate whether there is a bias (a tendency toward the same sign and value of departure from the official site between one's measurement of temperature and that of the official site), thereby arriving at a correction to the official predictions. This correction factor, which undoubtedly varies seasonally, can then be applied to future forecasts and may be especially valuable when predicted temperatures are near crucial values, such as freezing, 0°C (32°F).

An example of such a correction log is shown in Figure 6.9. Based on the temperature data presented in Figure 6.9, one might wish to decrease the officially predicted temperatures by a couple degrees Fahrenheit during the day and maybe a bit more at night in order to arrive at a more accurate prediction for one's own location. Such corrections, however, might vary from season to season and between sunny (or clear) and cloudy days (nights); they might have opposite signs between day and night in some seasons and the

Sample Data Spreadsheet					
Your Location		Official Location		Guidance Forecast (for Official Location)	
Max Temp	Min Temp	Max Temp	Min Temp	Max Temp	Min Temp
77	55	80	56	80	55
72	48	74	52	75	51
66	55	68	60	66	58

FIGURE 6.9. Daily record of maximum and minimum temperatures measured locally and as obtained from official guidance issued by the U S. weather service.

same sign in other seasons. It all depends on the effects of terrain and other factors affecting the local microclimate. It is best to make such comparisons over a long period of time and under differing weather conditions.

One can also apply the same principles to snowfall prediction. Snowfall amounts tend to be overpredicted in the smaller storms and underpredicted in the larger storms. By plotting a graph of predicted versus verified snowfall amounts over a long period of time (thereby acquiring many data points), one can gain an understanding of bias in the local snowfall predictions. Snowfall, as well as rainfall amounts, can vary locally from valley to hillside or from ocean to inland sites or from city to countryside.

Summing up

Weather at middle latitudes is driven by the exchange of heat and moisture between low and high latitudes. Warm air is constantly surging poleward and cold air equatorward. This movement of airstreams continually disturbs the tenuous balance between Coriolis and pressure gradient forces, such that the former must keep playing "catch up" with the latter. The result of this imbalance is patterns of divergence and convergence that allow cyclones and anticyclones to move and develop. We see this happening in ordinary surface weather maps in the form of cold- and warm-air advections, which help the forecaster identify areas where interesting weather events are taking place.

It is fortunate and highly satisfying that an essentially chaotic fluid such as the atmosphere can be understood at all and that its behavior can be predicted out to periods of more than a week. Equally satisfying is that our senses allow us to identify recognizable cloud types and cloud patterns that fit within conceptual models such as the comma-cloud model. Although this knowledge is useful to the professional forecaster, individuals without the information available from computer models can still predict the behavior of our chaotic atmosphere for at least a day or two in advance.

This book is not intended to explain the complexities of the science of computer weather forecasts. Rather, it has put forth some important and key points that will aid an amateur weather observer, professional forecaster, or beginning meteorology student in his or her pursuit of predictions.

Above all, we stress the importance of understanding what we observe in the various and ever-changing cloud patterns. In our effort to force the weather observer to look up, Chapter 3 presents a library of cloud types that are directly related to the interpretation of the comma-cloud model.

Thus, we should never forget the value of watching the sky as the first and last step in making a forecast. While we have not discussed the smell of the air and not much is written in the scientific literature about this subject, there are many anecdotes circulating among weather watchers about the way the air smelled as a portend of conditions. Certainly, the science of smell tells us that our noses can trigger memories as well as any of our other senses. We all know about the smell of rain and some even can sense subtle odor changes related to thunderstorm development. There is a difference to the aromas on mornings with and without dew, and a few people claim to be able to smell snow before it falls. Similar lore is told about the ability of an approaching storm to affect some people's joints. This is just another reason why the best forecasts originate with local meteorologists. It is simply a matter of tuning in to nature.

However, in the final analysis a person's eyes and intuition are some of the best tools for short-term weather predictions, and it is this sense that the present book hopes to foster. Unlike the automated surface weather observations systems (ASOS), one's vision can distinguish among cloud types, and one can view more of the celestial dome than any ceilometer (the device that measures the height of cloud bases). Indeed, it is a shame that, for practical reasons, cloud types are no longer plotted on weather maps.

Despite the volume of information available to the forecaster and the public, it is better to integrate observations within simple models or concepts rather than try to assimilate every scrap of information. We have emphasized the importance of understanding the basic cloud forms and where they occur within the classic cyclone model, such as how stratocumulus that cover the entire sky can be distinguished from the more stormy stratus clouds (by observing the sky through holes in the cloud deck), how changes in wind direction can signify changes in the weather or the passage of a front, and how the sky quadrant in which clouds seem to be thickening can signify the type of storm and its subsequent passage with respect to the observer. Because the human brain cannot account for all the data, simple models, such as the classic comma-cloud cyclone model, are essential for integrating observations into an understandable, conceptual pattern. Finally, it is important for the observer to recognize that not all data can fit neatly into a conceptual model, and to be aware when these deviations from the classic patterns might be occurring and to make the appropriate adjustment in one's thinking.

USEFUL METEOROLOGICAL WEB ADDRESSES

Weather maps (M); Forecast maps (FM); Station data (D); Radar imagery (R); Local station weather forecasts (F); Global weather maps (G); Upper-air maps (U); Health (H); Tropical (T); Satellite imagery (S); Severe weather (SW); Archived maps or data (A)

Unisys weather
- http://weather.unisys.com/
- M; FM; R; U; F; T; H; S

Weather Underground
- http://www.wunderground.com/ndfdimage/viewimage
- FM; R, SW, T; F; A, U

World maps
- http://www.meteo.psu.edu/~gadomski/WORLD/2010102012.gif
  (change date in url to access desired map)
- G; A, U

W S Radar Radar
- http://radar.weather.gov/ridge/radar.php?rid=ccx&product=NOR
- R

RAPS real time weather
- http://weather.rap.ucar.edu/model/
- FM

E-Wall weather, Penn State
- http://www.meteo.psu.edu/~gadomski/ewall.html
- F, D, A, U

AccuWeather
- http://www.accuweather.com/en/us/state-college-pa/16801/month/335315
- F, A

NOAA maps
- http://www.hpc.ncep.noaa.gov/dailywxmap/index_20101015.html
  (change date in url to access desired map)
- WM, U, A

WeatherShack
- http://www.weathershack.com/?gclid=CKqe47HYhbICFcRM4AoduXkA7Q
- Home Weather Stations

La Crosse Technology
- https://www.google.com/search?q=la+crosse+technology&ie=utf-8&oe=utf-8&aq=t&rls=org.mozilla:en-US:official&client=firefox-a
- Home Weather Stations

National Weather Service
- http://www.erh.noaa.gov/box/papers/blizzard78/mainblizzardof78.htm
- 1978 Snowstorm, referred to in text

INDEX

cumulonimbus (cb), 35, 110–14, *111, 113, 115*
cumulus congestus (cg), 35, 108–10, *109*
cumulus humilis, 107–8, *107, 109*
    formation of, 104, *108*
    towering cumulus (tc), 35, *90, 95*, 108–9,
        110, 114, 147, 155
    virga (precipitation trails) of, 105–6, *106*
    wind direction and, 187
cyclogenesis
    atmospheric instability and, *59*, 81–83
    of coastal storms, 70–77, *72, 74, 75*
    condensation warming and, 83–84
    energy sources for, 80–81
    fronts and, 74–75, 76, 124–25
    geographic areas favorable to, 84–85
    intensification and, 68–70, *69, 70*
    jet streak and, *75*, 76–77
    surface highs and lows and, 48–51, *49*
    temperature advection and, 42, 44–47,
        46, 54–55, 68–70, *69, 70*
    temperature inversions and, 57–60, *59*
    three-dimensional aspect of atmospheric
        circulation, 126–28, *127, 128*
    upper atmosphere and, 51–55, *54*
    vertical exchange and, 55–57, *56, 57, 59*,
        60–62
cyclones
    clouds and classic model of, *39, 41, 46*,
        49, 95, 119–20
    life cycle of, 77–80, *79*
    occluded fronts and, 65–68, *67*
    see also cyclogenesis
cyclonic motion, 15–16, 17, 19–23, *22*, 33
cyclostrophic balance, 18

Dalton, John, 124
daytime heating, 151–53, *153*
deposition, 90
dewpoint, 24
    situating of dewpoint sensor, 192
    in weather observation log, *191*, 191–92
divergence, 15, 17, 19, 30
    anticyclonic (clockwise) motion and,
        136, 138
    cyclogenesis and, 44, 73–74, 126–27, *128*

frictional divergence, 83–84, 86
    lake-effect snow, *149*, 150
dry convection, 31
dry tongue, *46*, 62, 66–67, *67*, 77, 182–83,
    189, *189*

embedded vortices, 52–53, *54*
ensemble weather forecasts, 165–72, *167, 168*,
    170, 171
explosive cyclogenesis (bombs), 68–70, 71
    geographic areas favorable to, 84–85

fog (fg), *95, 101*, 102–4, *103*
    dissipation of, 103, *103*
    mountain valley effects and, 156, 157
forecast funnel. *See* Snellman forecast funnel
fractus/fracto, as cloud name antecedent, 35
Franklin, Benjamin, 133
friction, lake-effect snow and, 149–50, *149*
friction layer, 18–19, 22
frictional convergence, 38, 85–86, *86*
frontogenesis, 74–75, 76
fronts, *39*, 40, 41, *41*, 50, 124–25
FROPA symbol, 50

geopotential height, 51
geostrophic balance, isobar spacing and, 17–20
geostrophic wind speed, 18
gradient, defined, 6
gravitational force, pressure gradients and,
    6–7
Great Lakes. *See* lake-effect snow
Gulf Stream, 20, 73, 77, 83

Hadley cell, 20–21
heat lows, 20, 124
Howard, Luke, 34
Hurricane Sandy, 166
hygroscopic (water attracting) nuclei, of
    clouds, 89

Icarus, 8
inertia, Coriolis force and, 11
interstices, in clouds, 97, 99
iridescence, in clouds, *99*, 99–100